LYMPHOKINES

Series editors

David Rickwood

Department of Biology, University of Essex, Wivenhoe Park,
Colchester, Essex CO4 3SQ, UK

David Male

Institute of Psychiatry, De Crespigny Park, Denmark Hill,
London SE5 8AF, UK

LYMPHOKINES

Anne S.Hamblin

Immunology Department, United Medical and Dental Schools of
Guys & St Thomas', St Thomas' Campus, Lambeth Palace Road,
London SE1 7EH, UK

OXFORD · WASHINGTON DC

IRL Press
Eynsham
Oxford
England

©1988 IRL Press Limited

First published 1988
Reprinted 1989, 1990

British Library Cataloguing in Publication Data

Hamblin, Anne S.
 Lymphokines.
 1. Man. Immune reactions. Role of lymphokines.
 I. Title   II. Series
 616.07'95

 ISBN 1 85221 055 9

Printed by Information Press Ltd, Oxford, England.

Preface

Lymphokines are a group of signalling molecules involved in communication between cells, particularly those of the immune system. Lymphokine-mediated events occur during the initiation and effector stages of immune responses and the development of haematopoietic cells. Consequently, an understanding of lymphokine biology is essential to understand how immune reactions develop and how they are controlled.

For many years the identity and characteristics of individual lymphokines were obscured by the difficulty in purifying them and the minute quantities which were active in biological assays. Within the last 3 years there has been an explosion of information on lymphokine structures and genes, along with a clarification of the roles of individual lymphokines. This was a consequence of developments in molecular biology, which has produced gene clones encoding many of these molecules.

This book focuses particularly on the structures and functions of these newly characterized lymphokines, and it brings together detailed information from many primary sources. Some lymphokines still await full characterization, as does the description of the lymphokine receptors. This is a particularly important field since the cellular response to a particular mediator depends not just on the expression of the mediator, but also on control of receptor expression as it occurs in individual cells. The possibilities that lymphokines are involved in the immunopathology of a number of diseases, and may provide new therapeutic modalities, provide the incentive for combined extensive investigation of these molecules.

A.S.Hamblin

Acknowledgements

I am grateful to Dr Claire Sharrock, Dr C.Sanderson, Dr P.Thorpe and members of the Department of Immunology at St Thomas' for their help with discussion or provision of articles. I am particularly thankful to NAS and KEHS for supporting me whilst I wrote in the attic.

Contents

Abbreviations

ADCC	antibody-dependent cell-mediated cytotoxicity
AIDS	acquired immunodeficiency syndrome
anti-TAC	anti-IL2 receptor antibody
APC	antigen-presenting cell
BCGF	B cell growth factor
BFU	blast forming unit
CFU	colony forming unit
CFU-E	CFU erythrocyte
CFU-Eo	CFU eosinophil
CFU-GEMM	CFU granulocyte, erythrocyte, monocyte, megakaryocyte
CFU-GM	CFU granulocyte, macrophage
CFU-Meg	CFU megakaryocyte
CSF	colony stimulating factor
DNP	dinitrophenol
E	erythropoietin
ELISA	enzyme-linked immunosorbent assay
f-Met-Leu-Phe	formyl-methionyl leucyl phenylalanine
G-CSF	granulocyte-CSF
GM-CSF	granulocyte-macrophage-CSF
HIV	human immunodeficiency virus
IFN	interferon
Ig	immunoglobulin
IL	interleukin
IL-2^r	IL-2 receptor
LAK	lymphokine activated killer cell
LPS	lipopolysaccharide
LT	lymphotoxin
M-CSF	macrophage-CSF
MHC	major histocompatibility complex
NK	natural killer
PMA	phorbol myristate acetate
RIA	radioimmunoassay
SAC	*Staphylococcus aureas* Cowan strain 1
Tc	cytotoxic T cells

TFR	transferrin receptor
T$_H$	helper T cells
TNF	tumour necrosis factor

1

Lymphokines

1. Discovery of lymphokines

The 1960s were a time of intense investigation into the origins and regulation of cellular immune responses. At that time, cell-mediated immunity to antigens was assessed *in vivo* by the production and cellular transfer of delayed-type hypersensitivity skin reactions, and *in vitro* by the transformation of sensitized lymphocytes by specific antigen and the inhibition of macrophage migration following interaction between lymphoid populations containing macrophages, sensitized lymphocytes and specific antigen. The demonstration that cell-free soluble factors generated *in vitro* in the culture supernatants of sensitized lymphocytes incubated with antigen, (i) could produce skin lesions similar to delayed-type hypersensitivity (1), (ii) were mitogenic for lymphocytes (2), and (iii) could cause macrophage migration inhibition (3), suggested that molecular mediators were involved in cellular immune responses (*Figure 1.1*).

The term lymphokine was introduced in 1969 to describe 'cell-free soluble factors generated by the interaction of sensitized lymphocytes with specific antigen and expressed without reference to the immunological specificity' (4). The generic term was chosen to emphasize their origins (lymphocytes) and also their role in the maintenance of the physiology of the immune system (kinesis). The evidence implicated thymus-derived T lymphocytes as the cells which interacted specifically with antigen to release the lymphokines which then acted non-specifically on target cells. Further work showed that crude culture supernatants could influence the *in vitro* behaviour of a large number of target cells in many different ways, and led to the view that many cellular immune interactions were regulated by soluble factors (5).

During the 1970s the term lymphokine became more widely used, not only to describe the large number of biological activities of antigen-activated T lymphocytes presumed, but not proven, to be ascribable to biochemical factors in the culture supernatants, but also those in culture supernatants of lymphocytes non-specifically activated with mitogens (e.g. phytohaemagglutin and concan-

1

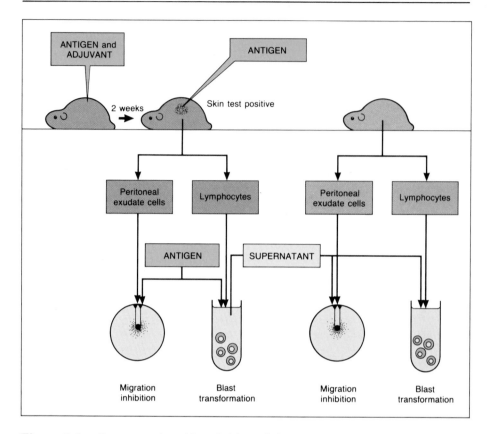

Figure 1.1. Demonstration of lymphokines. Guinea pigs are injected with antigen in complete Freund's adjuvant: after 2 weeks the animals give a positive delayed hypersensitivity skin test to the antigen, their peritoneal exudate cells show inhibited migration in the presence of antigen and their lymphocytes proliferate in response to antigen. These effects can be reproduced on cells from untreated guinea pigs, using tissue culture supernatants of antigen-activated lymphocytes from the immunized animals.

avalin A) or cultured cell lines of both lymphoid as well as non-lymphoid origin. Similar activities were also found in body fluids such as serum and urine. During this period there was considerable scepticism regarding the role and even existence of lymphokines. For many workers, the large number of activities [more than 100 were cited in a list published in 1979 (5)], their disparate sources and their lack of biochemical characterization were enough to suggest that the whole concept was based on artefact.

However, by the late 1970s, a number of technical advances had led to the purification to homogeneity of some of the lymphokines. This progress has continued with the introduction of molecular biology techniques, resulting in a better understanding of the structure of these mediators and their function.

It is now clear that cells of the immune system secrete and respond to soluble factors which exert wide ranging effects. Whilst T cells and macrophages are a major source, other cells can also produce them. The factors are not only important in the regulation and differentiation of cells responding to antigen, but also the inflammatory and physiological interactions between immune and non-immune cells.

2. Nomenclature and classification

At first, lymphokines were named on the basis of the activity they produced *in vivo*, or more frequently *in vitro*. Their names were abbreviated to acronyms. For example, migration inhibitor factor, or MIF, was generated in cultures of antigen-activated lymphocytes and, when added to non-immune peritoneal macrophages, inhibited their migration from capillary tubes (3,5).

Other terms were introduced in attempts to order the enormous number of acronyms and titles applied to substances under the general term lymphokine. Thus 'monokine' and 'cytokine' were used to denote the fact that monocytes were the source of some biological mediators and that non-lymphoid cells produced others, and to differentiate them from lymphokines produced by lymphocytes (6). The term interleukin (between leukocytes) first appeared in 1979 to 'free nomenclature from the constraints associated with definitions by single bioassays' (7). Constraining bioassay-based names were felt to be inappropriate when it became clear that a variety of biological activities were actually different effects of the same substance. For example, the factor derived from monocytes causing lymphocyte activation, which had previously been known as lymphocyte activating factor (LAF) appeared the same as that previously known as mitogenic protein (MP), T cell-replacing factor III (TRF-III), a B cell activating factor (BAF) and a B cell differentiation factor (BDF). The factor was renamed interleukin 1 (IL-1). At the same time it was realized that T cell growth factor (TCGF) was the same as thymocyte mitogenic factor (TMF) and killer cell helper factor (KHF) and this was renamed IL-2 (7). The terms interleukin, monokine and cytokine are now used widely, although the term lymphokine has tended to retain its position as an umbrella term to describe soluble proteins which influence cells of the immune system regardless of their source.

The attribution of a variety of biological effects to the same substance, as described above for IL-1 and IL-2, was made possible by improved biochemical analysis of cell supernatants allowing purification of factors to homogeneity and then production of large amounts by genetic engineering. Currently, the cloned lymphokines include six interleukins as well as a number of other lymphokines which retain their activity-based names (*Table 1.1*). Whilst there is every reason to believe that many of the 100 activities described in 1979 are caused by the same factors, the actual number of lymphokines remains to be determined. In this book the emphasis is on the structure and biological activity of those lymphokines which have been cloned. There are certainly other important lymphokines which have yet to be cloned, and there are several lymphokines which have been

Table 1.1.   Cloned lymphokines and their alternative names

	Acronym	Alternative title
Interferon-γ	IFN-γ	
Interleukin 1α	IL-1α	Lymphocyte activating factor (LAF); mitogenic protein (MP); T cell replacing factor III
Interleukin 1β	IL-1β	(TRF-III); B cell activating factor (BAF); B cell differentiation factor (BDF); endogenous pyrogen (EP); leukocyte endogenous mediator (LEM); serum amyloid A (SAA) inducer; proteolysis inducing factor (PIF); catabolin; haematopoietin 1 (HP1); mononuclear cell factor (MCF)
Interleukin 2	IL-2	T cell growth factor (TCGF); thymocyte mitogenic factor (TMF); killer cell helper factor (KHF)
Interleukin 3	IL-3	Multi-potential colony stimulating factor (multi-CSF); burst promoting activity (BP); haemopoietic cell growth factor (HPGF); persisting cell stimulating factor (PSF); mast cell growth factor (MCGF); haematopoietin 2 (HP2)
Interleukin 4	IL-4	B cell stimulation factor 1 (BSF-1); T cell growth factor II (TCGF-II); mast cell growth factor II (MCGF-II)
Interleukin 5	IL-5	T cell replacing factor (TRF), B cell growth factor II (BCGF-II); eosinophil differentiation factor (EDF)
Interleukin 6	IL-6	Interferon β_2 (IFN-β_2); B cell stimulation factor 2 (BSF-2); B cell differentiation factor (BCDF); hybridoma/plasmacytoma growth factor (HPGF); hepatocyte stimulating factor (HSF)
Granulocyte-macrophage-colony stimulating factor	GM-CSF	Colony stimulating factor α (CSF-α); pluripoietin; neutrophil inhibition factor (NIF-T)
Macrophage-colony stimulating factor	M-CSF	Colony stimulating factor 1 (CSF-1)
Granulocyte-colony stimulating factor	G-CSF	Colony stimulating factor β (CSF-β)
Tumour necrosis factor	TNF	Cachectin; tumour necrosis factor α (TNF-α)
Lymphotoxin	LT	Tumour necrosis factor β (TNF-β)

partially characterized. The breadth of this book does not permit full description of these and the reader is referred to the general reading list for further information.

3. General aspects of lymphokines and lymphokine receptor structures and genetics

Early studies indicated that lymphokines were proteins or glycoproteins which were not immunoglobulins (Igs). In spite of many attempts, further consistent biochemical analysis proved more difficult. In the first place, only small amounts are made by cells in culture and therefore large cultures were needed to provide sufficient starting material for analysis. In the second place, activity was frequently lost on purification. Where information was acquired, it was often conflicting, particularly with respect to molecular size (8).

Several technological advances provided better means to investigate the biochemical structure of lymphokines. Cell lines and hybridomas could be grown in large volumes of culture medium from which the lymphokine could be purified (9). The cells could often be stimulated with a cocktail of agents to superinduce production. By both these means, the amount of starting material for purification could be increased. As biochemical separation and detection methods for proteins became more sensitive, it became possible to purify lymphokines to homogeneity from native sources (see Chapter 2). With the advent of gene cloning the means to produce very large amounts of protein in prokaryotes or eukaryotes were established. The availability of large amounts of homogeneous protein has permitted better analysis of lymphokine biological activity *in vivo* and *in vitro*, as well as providing potential therapeutic agents for clinical trial.

Lymphokines are encoded by genes for which there is only one copy per haploid cell. In common with most eukaryotic genes, the lymphokine genes are segmented, being composed of exons which are complementary to sequences in the mature mRNA and which are separated by introns which are not found in the mRNA. Most lymphokine genes consist of 3 – 4 introns and 4 – 5 exons and are located on a number of different chromosomes. It is of interest that in the human a number are located on chromosome 5 and that tumour necrosis factor (TNF) and lymphotoxin (LT) are closely associated on chromosome 6 within the major histocompatibility complex (see Chapter 3). The location of genes on the same chromosome raises the possibility that they may be closely linked and under the influence of common regulatory elements.

Lymphokines are usually produced by cells in response to induction signals generated from the cell surface. Many, for example granulocyte-macrophage-colony stimulating factor (GM-CSF), IL-2, IL-3 and interferon γ (IFN-γ), have common nucleotide sequences in the 5' flanking region of their genes which may be important in the initiation of transcription and indeed in the coordinate expression of the genes so often seen on cell stimulation. The production of lymphokines seems to be controlled, particularly at the level of transcription.

Detailed structure of lymphokine genes and their flanking sequences is therefore providing information which will unravel control of their regulation at the DNA level.

cDNAs for lymphokines predict mature proteins of around 100–200 amino acids. Most have a clearly defined hydrophobic signal sequence of around 20 amino acids, which is cleaved to give the mature protein. It is notable that for those factors derived from macrophages (TNF, IL-1α and IL-1β) a pre-sequence is cleaved from the mature protein which is unusually long (70 or more amino acids). Since the signal sequences are associated with transport of the proteins out of the cell, this difference in size may represent an important difference in secretory mechanisms between macrophages and lymphocytes.

Amino acid sequence data show that many lymphokines contain cysteine residues which may be important in the formation of intramolecular disulphide bonds. Using site-directed mutagenesis and chemical reduction the importance of these disulphide bridges for the tertiary structure of the molecule and its biological activity has been determined for several of the lymphokines.

Molecular weight determinations of lymphokines purified from native sources frequently suggest that they are not homogeneous and that they have larger weights than that predicted from the cloned genes. The discrepancies between predicted and observed molecular weights arise from post-translational modification, particularly glycosylation. Most lymphokines are variably glycosylated (usually N-glycosylated) and the glycosylated proteins may form oligomers. Size estimations of natural proteins suggest a range of molecular weights which are reduced by separation under reducing conditions on SDS–PAGE. Further reductions in molecular weight heterogeneity may be achieved by incubation of the lymphokine with glycosidases which cleave off glycosylated side chains. The function of the glycosylation is unclear since recombinant products made in *Escherichia coli*, which are not able to glycosylate proteins, often have the same biological activities and the same half-lives *in vivo* as their glycosylated equivalents. Furthermore, site-directed mutagenesis of potential glycosylation sites does not alter biological function.

Most molecular genetics to date has been undertaken on mouse and human proteins with occasional information on other species. Comparison of both cDNA and genomic sequences of different lymphokines in one species and the same lymphokine beween species has been used to determine possible relationships between lymphokines as well as phylogenetic conservation of structure. Thus, IL-1α and IL-1β are structurally related and have the same or very similar biological activities suggesting they may have arisen from gene duplication. IL-1α shows 61–65% amino acid homology between human, rabbit and mouse and IL-1β shows 27–33% homology with IL-1α in the three species. These results suggest that the IL-1 genes arose from duplication before or during vertebrate evolution and then diverged independently (10). Homology between mouse and human lymphokines ranges from close [e.g. IL-5 (70%)] to very poor [e.g. IL-3 (29%)] (11).

Following from characterization of the structure of lymphokines comes the characterization of their receptors on cell surfaces. So far, detailed information

Table 1.2. Bioassays for lymphokines[a]

IFN-γ	Virus yield reduction
	Inhibition of viral RNA and protein synthesis
	Cell growth inhibition
IL-1	Co-mitogenic stimulation of mouse thymocytes in presence of mitogen
	Proliferation of D10.G4 T cell line
	Production of IL-2 by LBRM-33 1A5 line
	Production of fibroblast prostaglandin
	Bone resorbtion
	Proteoglycan release from cartilage
IL-2	Proliferation of IL-2-dependent T cell lines
IL-3	Haematopoietic colony formation from bone marrow
	Proliferation of IL-3-dependent lines
	Induction of 20α-steroid dehydrogenase in nude mouse spleen cells or bone marrow
IL-4	Co-stimulation of mouse splenic B cells in presence of anti-Ig
	Increase in class II MHC expression by B cells
	Induction of IgG, by T-depleted spleen cells
	Induction of IgE by T-depleted spleen cells
	Proliferation of mast cell lines
	Proliferation of T cell lines
IL-5	Induction of proliferation and IgM production by B cell lines (mouse)[b]
	Proliferation or antibody secretion by large B cells (mouse)[b]
	Induction of secondary anti-DNP IgG antibody by DNP-primed T-depleted B cells (mouse)[b]
	Differentiation of eosinophils in bone marrow cultures
IL-6	Ig production by normal B cells (human)[b]
	Ig production by Epstein–Barr virus-transformed cell lines (human)[b]
GM-CSF G-CSF M-CSF	Haematopoietic colony formation of bone marrow cells
TNF LT	Antiproliferative activity on certain tumour cell lines

[a]Ref. (14) and Clemens et al. (Further reading).
[b]Indicates that assay only available for that species.

is only available for the IL-2 receptor of which one chain has been molecularly cloned and the other characterized by gel electrophoresis and binding studies (see Chapter 2). The final structure and its interaction with IL-2 have yet to be fully determined. For most other lymphokines the density and affinity of receptors have been investigated using radiolabelled pure proteins. Like the lymphokines themselves, receptor expression may be induced following stimulation of the cells. Thus, receptors which are normally absent or present in low numbers on cells increase in number during cell activation and then return to resting levels.

It may be anticipated that receptors for all lymphokines, like other ligands, will consist of extracellular domains which bind the ligand, a hydrophobic transmembrane region and an intracellular domain. The events that take place following binding of ligand to receptor have been characterized in other systems and include synthesis of cyclic nucleotides and hydrolysis of phosphatidylinositol 4,5-bisphosphate, as well as activation of protein kinase C and elevation of intra-cytoplasmic calcium (12). The extent to which these mechanisms for generating second messengers occur when lymphokines bind to receptors is of great interest, but at the moment is poorly understood. A final, further general feature of receptor – ligand interactions is the internalization of the complex by receptor-mediated endocytosis (13). Many studies suggest that lymphokines along with other ligands enter the cell in this fashion, although the final consequences of this and the fate of the two components is as yet undetermined.

4. General aspects of lymphokine biology

Much of the information available about lymphokine biology comes from studies of their generation and assay *in vitro*. There have been less analyses of the effects of injection of lymphokines into animals although the recent clinical use of cloned factors is providing a surge of new data (Chapter 5). There are even fewer descriptions of lymphokine generation and lymphokine receptor distribution *in vivo*. It follows that our understanding of lymphokine biology is often constrained by the limitations of interpreting *in vitro* data and these are worth considering before passing to a discussion of general aspects of lymphokine biology.

Detection of lymphokines is usually by *in vitro* bioassay (14), although assays based on the antigenicity of molecules [e.g. radioimmunoassay (RIA) or enzyme-linked immunosorbent assays (ELISA)] are becoming more available. A list of the common bioassays for lymphokines is shown in *Table 1.2*. This is by no means exhaustive but exemplifies the range of assays which are used. Bioassays for lymphokines have rarely involved the effect of a single factor on a single cell type, although in recent years it has been possible to assay the effect of a recombinant lymphokine on a cell line and use monoclonal antibodies in tests to produce assays with restricted variability (15). However, where mixtures of lymphokines, as found in cultures of activated lymphocytes, are assayed, or where impure target cells are used, there may be difficulty in interpreting the results. These problems arise because several lymphokines can produce similar biological effects. Thus, IL-4 has been found to cause stimulation of thymocytes or T cell clones (16) used in assays for IL-1 and IL-2, respectively, and TNF and LT are indistinguishable biologically (see Chapter 4). Furthermore, a lymphokine may induce the production of other lymphokines which may influence different cells in a mixed target cell population, or which may interfere positively or negatively with assays on homogeneous cells. Induction of one lymphokine by another has often been described, for example IL-1 and TNF stimulate fibroblasts to produce

Table 1.3. Cellular sources and targets of lymphokines

Lymphokine	Cellular source	Cellular target
IFN-γ	T cells, natural killer (NK) cells	Macrophages, T cells, B cells, NK cells
IL-1α and β	Macrophages, endothelial cells, large granular lymphocytes, B cells, fibroblasts, epithelial cells, astrocytes, keratinocytes, osteoblasts	Thymocytes, neutrophils, hepatocytes, chondrocytes, muscle cells, endothelial cells, epidermal cells, osteocytes, macrophages, T cells, B cells, fibroblasts
IL-2	T cells	T cells, B cells, macrophages
IL-3	T cells	Multipotential stem cells, mast cells
IL-4	T cells	T cells, mast cells, B cells, macrophages, haematopoietic progenitors
IL-5	T cells (mouse)	Eosinophils, B cells (mouse)
IL-6	Fibroblasts, T cells	B cells, thymocytes
GM-CSF	T cells, endothelial cells, fibroblasts, macrophages	Multipotential stem cells
M-CSF	Fibroblasts, monocytes, endothelial cells	Multipotential stem cells
G-CSF	Macrophages, fibroblasts	Multipotential stem cells
TNF-α	Macrophages, T cells, thymocytes, B cells, NK cells	Tumour cells, transformed cell lines, fibroblasts, macrophages, osteoclasts, neutrophils, adipocytes, eosinophils, endothelial cells, chondrocytes, hepatocytes
LT	T cells	Tumour cells, transformed cell lines, neutrophils, osteoclasts

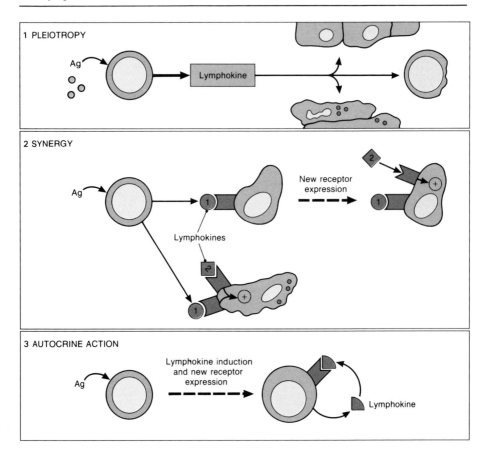

Figure 1.2. Modes of lymphokine action. Pleiotropy: a single lymphokine produces different effects by acting on different cell types (**1**). Synergy is seen when one lymphokine induces the receptor for another lymphokine or when both lymphokines are required together to stimulate a cell (**2**). Some lymphokines act on the cell which produced them in an autocrine fashion (**3**).

IL-6 (17) and IL-1 stimulates T cells to produce a number of lymphokines (see Chapter 3).

Lymphokine bioassays frequently reflect only one aspect of a particular lymphokine's biological activity. Where the effect of a lymphokine is restricted, such *in vitro* assays reflect the important role played by that lymphokine *in vivo*. Thus, IL-2 is usually bioassayed by its ability to cause proliferation and growth of a T cell line which constitutively carries the IL-2 receptor (e.g. CTLL line) and this mirrors the important role that IL-2 plays in the clonal expansion of T cells *in vivo*. Other common bioassays reflect only one of many activities which the lymphokine has. Thus, IL-1 is assayed for its co-stimulatory effects with mitogen on mouse thymocytes. It can be shown to exert many other biological effects and the relevance of the mouse thymocyte assay to the major functional properties of IL-1 remains to be determined (see Chapter 4). The reader is urged to consider these points when evaluating the reported *in vitro* biological effects of a lymphokine. However, from working with cloned lymphokines and cell lines the following features may be deduced.

Lymphokines are made by different cells (*Table 1.3*) often in response to their activation. Some, like IL-2, seem to be made only by a restricted type of cell (i.e. T cells) whilst others, like IL-1, can be produced by very different cell types. Whilst there are many cell lines which produce lymphokines *in vitro* constitutively or after activation (18), the normal sources of any lymphokine *in vivo* may not be clear. This particularly applies to the colony stimulating factors (CSFs) where the natural source in the bone marrow stroma is not known.

The target cells of lymphokine action may also be restricted or of very diverse types (*Table 1.3*). At early stages in lymphokine research it was suspected that lymphokines might have multiple activities on different target cells (5). The availability of recombinant lymphokines has confirmed that they are often pleiotropic. Thus, IL-1 acts on T cells and is considered important in the immune response to antigens. It also produces inflammatory effects on a variety of target cells (19). It is a puzzle how one molecule can exert such very different effects on different target cells. It also remains to be seen whether all of these targets are affected *in vivo* and under what circumstances.

An important aspect of lymphokine activity is that they frequently work together, with each other or with another stimulant, to produce effects. Thus, for example, LT and IFN-γ have a potent synergistic effect in antiproliferative assays *in vitro* and antitumour effects *in vivo* (20) and IL-2 and IL-4 synergize to cause proliferation of T cell clones (16). Such synergistic action is not understood. However, it may be that the interaction of one lymphokine with its receptor may prime a cell to become responsive to a second signal. Alternatively, the occupancy of two receptors may deliver the correct signal to allow the cell to respond (*Figure 1.2*). A better understanding of lymphokine – receptor intracellular signalling should lead to a further understanding of how these synergistic interactions arise and their function(s).

Lymphokines have been frequently referred to as hormones (21,22), in that they are produced by one cell and may act at a distant site. The extent to which lymphokines affect cells which are not local to their release is important in determining their role in normal physiology and pathology. Not only do lymphokines act as hormones but they also act in an autocrine fashion back on the cell which produces them. IL-1 and IL-2 are produced by macrophages and T cells, respectively, and can bind to a receptor on the same cell to produce the desired effect. Here again, the extent to which such autocrine action occurs *in vivo* is important in considering the maintenance of an immune response.

5. Further reading

Lymphokines (1980 – 1987) Academic Press, Orlando, Vols 1 – 14.
Clemens,M.J., Morris,A.G. and Gearing,A.J.H. (eds) (1987) *Lymphokines and Inter-ferons—A Practical Approach*. IRL Press, Oxford.

Cohen,S., Pick,E. and Oppenheim,J.J. (eds) (1979) *Biology of the Lymphokines.* Academic Press, New York.

Hadden,J.W. and Stewart,W.E. (eds) (1981) *The Lymphokines: Biochemistry and Biological Activity.* Humana Press, New Jersey.

Sorg,C. and Schimpl,A. (eds) (1985) *Cellular and Molecular Biology of Lymphokines.* Academic Press, New York.

Webb,D.R., Pierce,C.W. and Cohen,S. (eds) (1987) *Molecular Basis of Lymphokine Action.* Humana Press, New Jersey.

6. References

1. Bennett,B. and Bloom,B.R. (1968) *Proc. Natl. Acad. Sci. USA,* **59**, 756.
2. Kasakura,S. and Lowenstein,L. (1970) *J. Immunol.,* **105**, 1162.
3. Bloom,B.R. and Bennett,B. (1966) *Science,* **153**, 180.
4. Dumonde,D.C., Wolstencroft,R.A., Panayi,G.S., Mathew,M., Morley,J. and Howson,W.T. (1969) *Nature,* **224**, 38.
5. Waksman,B.H. (1979) In Cohen,S., Pick,E. and Oppenheim,J.J. (eds), *Biology of the Lymphokines.* Academic Press, New York, p. 585.
6. Cohen,S. *et al.* (1977) *Cell. Immunol.,* **33**, 233.
7. Aarden,L.A. *et al.* (1979) *J. Immunol.,* **123**, 2928.
8. Yoshida,T. (1979) In Cohen,S., Pick,E. and Oppenheim,J.J. (eds), *Biology of the Lymphokines.* Academic Press, New York, p. 259.
9. Kappler,J. and Marrack,P. (1986) In Weir,M. (ed.), *Handbook of Experimental Immunology.* Blackwell Scientific Publications, Oxford, p. 59.1.
10. Lomedico,P.T., Gubler,U. and Mizel,S.B. (1987) *Lymphokines,* **13**, 139.
11. Sanderson,C.J., Campbell,H.D. and Young,I.G. (1988) *Immunol. Rev.,* **102**, 29.
12. Poste,G. and Crooke,S.T. (eds) (1985) *Mechanisms of Receptor Regulation.* Plenum Press, New York.
13. Brodsky,F.M. (1984) *Immunol. Today,* **5**, 350.
14. Hamblin,A.S. and O'Garra,A. (1987) In Klaus,G.G.B. (ed.), *Lymphocytes – A Practical Approach.* IRL Press, Oxford, p. 209.
15. Cherwinski,H.M., Schumacher,J.H., Brown,K.D. and Mosmann,T.R. (1987) *J. Exp. Med.,* **166**, 1229.
16. Spits,H., Yssel,H., Takebe,Y., Arai,N., Yokota,T., Lee,F., Arai,K., Banchereau,J. and de Vries,J.E. (1987) *J. Immunol.,* **139**, 1142.
17. Van-Damme,J., Opdenakker,G., Simpson,R.J., Rubira,M.R., Cayphas,S., Vink,A., Billiau,A. and Van Snick,J. (1987) *J. Exp. Med.,* **165**, 914.
18. Kelso,A., Glasebrook,A.L., Kanagawa,O. and Brunner,K.T. (1982) *J. Immunol.,* **129**, 550.
19. Oppenheim,J.J., Kovacs,E.J., Matsushima,K. and Durum,S.K. (1986) *Immunol. Today,* **7**, 45.
20. Stone-Wolff,D.S., Yip,Y.K., Kelker,H.C., Lee,J., Henriksen-De Stefano,D., Rubin,B.Y., Rinderknecht,E., Aggarwal,B.B. and Vilcek,J. (1984) *J. Exp. Med.,* **159**, 828.
21. Duram,S.K., Schmidt,J.A. and Oppenheim,J.J. (1985) *Annu. Rev. Immunol.,* **3**, 263.
22. Smith,K.A. (1984) *Annu. Rev. Immunol.,* **2**, 319.

2

Biochemistry of lymphokines

1. Preparation

1.1 Preparation from natural sources

A general scheme for the preparation of lymphokines from natural sources is shown in *Figure 2.1*.

Large volumes of culture supernatant may be obtained by large-scale cultures of cells which are known to produce the desired activity. The cells used may be peripheral blood mononuclear cells, but more frequently are hybridomas or cell lines from both lymphoid and non-lymphoid sources. To obtain a maximum yield of lymphokine the cells are superinduced with appropriate stimulants such as mitogens, phorbol esters or mixtures of these. Since lymphokines are secreted rapidly into the medium, this phase may be no more than 24 h and is frequently undertaken in serum-free medium to eliminate the problem of serum proteins contaminating subsequent purification steps. The culture supernatants are concentrated and then the proteins purified by a number of techniques separately or sequentially. The presence of the lymphokine is determined in the fractions obtained by bioassay or, if polyclonal or monoclonal antibodies to the lympho-kines are available, radioimmunoassay (RIA) or enzyme-linked immunosorbent assay (ELISA). Having established which protein fraction is bioactive, the protein is further characterized for molecular weight, homogeneity and pI. The amino acid composition and sequence of the homogeneous protein can then follow. This labour intensive procedure has been used for a number of lymphokines, an example of which is shown in *Figure 2.2*.

1.2 Preparation by recombinant DNA technology

The above purification schemes as regular sources of lymphokines have mostly been replaced by recombinant DNA methods. The relevant gene is inserted into prokaryotes, such as *Escherichia coli*, or eukaryotes, such as yeast, which replicate rapidly. Since these sources can provide many thousand times the

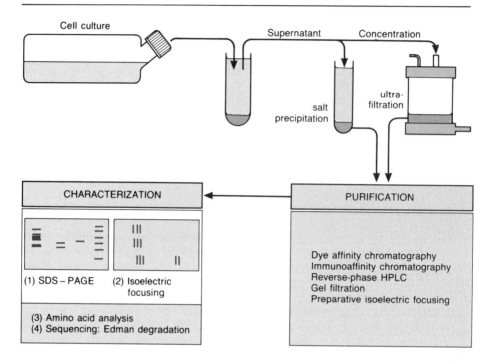

Figure 2.1. Purification and characterization of lymphokines from natural sources. The supernatant of a cell line producing the lymphokine is obtained and the protein concentrated by salt precipitation or ultrafiltration. The lymphokine is purified by successive fractionation methods, and the active fraction identified by RIA or bioassay at each stage. Analytical gels are used to characterize the purified lymphokine.

amounts obtainable by conventional means and can be cultured continuously they provide an excellent source of lymphokines. The lymphokine is often the major protein produced and its separation from vector proteins and contaminants is relatively straightforward.

There are two common approaches to isolating gene sequences for cloning, examples of which are shown in *Figures 2.3* and *2.4*. In the first, oligonucleotides are prepared to match a portion of the amino acid sequence of the protein and these are used to probe cDNA libraries. This requires purification and partial sequencing of the protein by the methods described above. In the second approach, protein sequencing is circumvented by isolation of mRNA from an appropriate cell. This is then translated *in vitro*, or more usually in *Xenopus* oocytes, and the translation product is detected by bioassay or by binding to an antibody if one is available. The fraction of mRNA which synthesizes the protein is then used to construct a cDNA library and the library probed (often with the mRNA) to identify a clone making a full-length cDNA insert. A particular feature of lymphokines which has been exploited in these studies is the fact that mRNA expression is usually induced by stimulation of cells. The RNA of interest may therefore be identified by comparison and subtraction of that from an

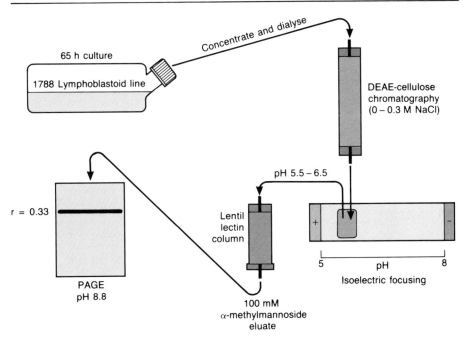

Figure 2.2.   Purification of human LT according to the method of Aggarwal *et al.*
(44). The lymphokine-containing supernatant fraction was concentrated, dialysed, and
purified by (i) ion-exchange chromatography, (ii) isoelectric focusing, (iii) lentil lectin
affinity chromatography and (iv) preparative gel electrophoresis. The resulting isolate
contained 10.7% of the original activity and was enriched 20 000-fold.

uninduced clone (plus – minus or subtractive hybridization). Once a double-
stranded cDNA has been prepared it can be inserted via a plasmid into bacteria
or yeast which replicate producing large amounts of protein.

   Cloned lymphokine genes and proteins can be sequenced and investigated in
various ways. Site-directed mutagenesis may be used to alter single amino acids
so that they no longer glycosylate or form disulphide bridges. Alternatively
lymphokine genes may be engineered into yeast or mammalian cells which will
glycosylate proteins to produce a more 'natural' product. In these ways the
biological importance of glycosylation and disulphide bonds can be studied.
Synthetic or cDNAs are used to probe a human genomic library to determine
the structure of the gene. The chromosomal location of the gene is determined
by hybridization to DNA from somatic cell hybrids containing known sets of
chromosomes from the species under investigation.

2.   Structure of lymphokines and lymphokine receptors

2.1 Interferon γ

Interferon γ (IFN-γ) is an antiviral and antiproliferative protein which also has
potent immunoregulatory effects on a variety of cells. These include activation

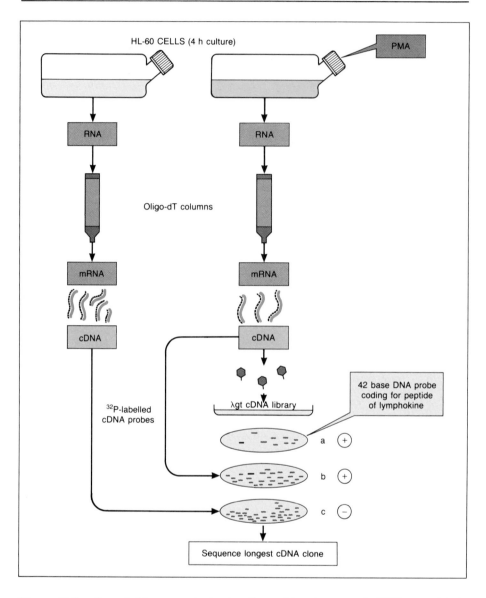

Figure 2.3.   Lymphokine gene cloning by oligoprobing, for example TNF (48). mRNA was isolated from PMA-stimulated (right) and unstimulated cells on oligo-dT columns. A cDNA library was prepared from the stimulated cell RNA. The colonies were hybridized to a DNA probe encoding a known peptide sequence of the lymphokine (a). Nine colonies which hybridized to this probe and to cDNA from the activated cells (b) but not the unstimulated cells (c) were identified and the longest cDNA insert was sequenced.

of macrophages, induction of class I and class II major histocompatibility complex (MHC) gene products on both macrophages and other cells of non-haematopoietic origin, and activation and differentiation of other immune cells. It is produced

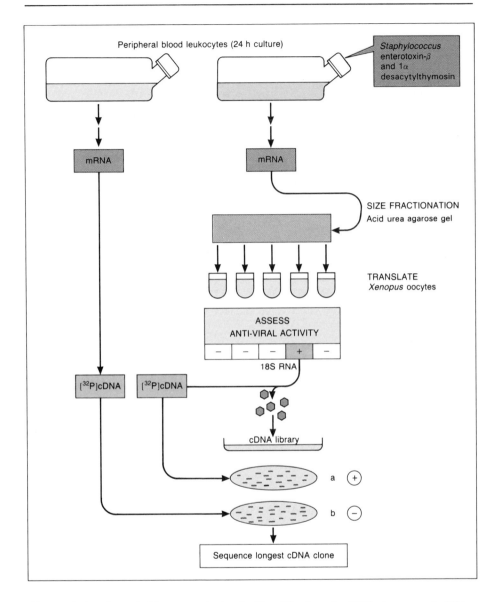

Figure 2.4.   Lymphokine gene cloning by identification of mRNA, for example IFN-γ (5). mRNA was isolated from stimulated leukocytes and fractionated. The fractions were translated in *Xenopus* oocytes and the translation products assessed for their ability to induce virus resistance. The active mRNA was used to make a cDNA library and 20 clones were isolated which hybridized to cDNA of stimulated cells (a) but not unstimulated cells (b). The longest cDNA insert was then sequenced.

by T lymphocytes from blood or lymphoid tissues upon stimulation with specific antigens, mitogens or alloantigens. Both CD4$^+$ and CD8$^+$ lymphocytes can produce IFN-γ, although the former are considered the major producers in

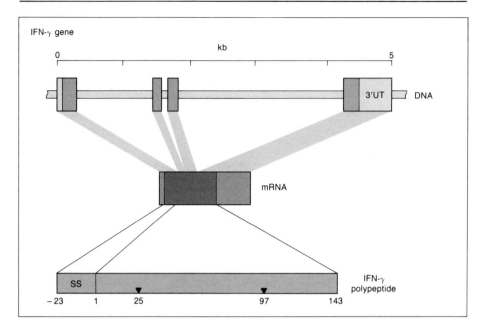

Figure 2.5. Human IFN-γ gene structure. The IFN-γ gene consists of four exons with 5' and 3' untranslated segments (lighter tone). The primary transcript is spliced to form mRNA which is translated to make a polypeptide of 166 amino acids of which the first 23 act as a signal sequence (SS) and are cleaved from the mature protein. N-Glycosylation sites are present at residues 25 and 97.

response to antigens (1). Original size fractionation of human IFN-γ produced by mitogen-stimulation of peripheral blood leukocytes was by molecular sieving and suggested a size of 35 – 70 kd. Subsequent purification to homogeneity using SDS – PAGE suggested two active forms of the glycosylated protein with molecular weights of 20 and 25 kd and pIs (using isoelectric focusing) of 8.3 and 8.5 (2). These two forms seem to be the product of a single gene [on chromosome 12 in the human (3) and 10 in the mouse (4)] with size and charge differences arising from differences in glycosylation. In their natural form, these species aggregate giving rise to the high molecular weight products seen in fractionation by molecular sieving.

The human cDNA coded for 166 amino acids of which the first 23 had characteristics of eukaryotic signal sequences in that they were mostly hydrophobic (5). The secreted protein consists of 143 amino acids and has a structure unrelated to that of IFN-α or -β suggesting that it has independent evolutionary origins. Amongst the various other species examined, there is little sequence homology at the amino acid level, which probably explains the lack of cross-species activity. Between man and mouse, homology is 40% and the mouse molecule has 10 fewer amino acids.

The structure of the DNA, spliced mRNA and polypeptide are shown in *Figure 2.5*. The 20 and 25 kd molecular weight natural forms of IFN-γ arise from glycosylation at residue 25 (both forms) and 97 (25 kd form) of the polypeptide sequence. The extent of glycosylation is variable but biological activity is unaffected by this; recombinant IFN-γ prepared in *E.coli* is biologically active.

A specific receptor for IFN-γ has been characterized which is distinct from the receptor for IFN-α and -β (6). Since many cells respond to IFN-γ it may be inferred that the receptor is widely distributed on different cell types and to date all human cells tested have been shown to bear receptors. The receptor is susceptible to the action of proteases and is therefore a protein. The molecular weight of the receptor on monocytes and haematopoietic cell lines is 140 kd whilst other cell types carry a receptor of molecular weight 95 kd. Molecular characterization of these structures is still awaited.

Binding of IFN-γ to its receptor leads to rapid internalization of the complex at 37°C by receptor-mediated endocytosis. It also results in the induction of several proteins associated with the antiviral state and immunoregulation. There is very little information concerning the events which occur between IFN-γ receptor interaction and activation of these genes. It is presumed this is via a second message which acts on the interferon-responsive sequences in the non-coding portion of the genes leading to their transcription (7).

2.2 Interleukin 1

Interleukin 1 (IL-1) is a polypeptide with diverse roles in immunity and inflammation (see Chapters 3 and 4). It is synthesized by many cell types but particularly monocytes and macrophages which have been activated. These cells are potent producers of IL-1 upon stimulation with a variety of agents including endotoxin, muramyl dipeptide, phorbol myristate acetate (PMA) and silica. Unstimulated monocytes possess low levels of message for IL-1 which rise 2 h after stimulation, followed at 3 h by IL-1 protein which is detectable outside the cell (8).

Early work suggested that many of the bioactivities in *Table 1.1* co-purified with a molecule of molecular weight 15–17 kd (9). Further characterization of natural IL-1 has demonstrated biologically active molecules of higher and lower molecular weight in culture supernatants and body fluids (10). In addition, IL-1 shows charge heterogeneity with a major species having a pI of 7 and minor species with pIs of 5 and 6. The initial confusion which arose from different results in different laboratories has been somewhat clarified by gene cloning. Several methods have been used to clone the genes for IL-1 including both oligo-probing and RNA selection and these have revealed that there are two related gene products, IL-1α and IL-1β (11). Most human IL-1 produced by stimulated macrophages is IL-1β, corresponding to the species of pI 6.8. IL-1α corresponds to the minor IL-1 species at pIs 5 and 6. The genes code for precursor proteins of 271 (IL-1α) and 269 (IL-1β) amino acids. Biologically active IL-1 can be derived from clones coding for the carboxy-terminal 154 amino acids (IL-1α) and 153 amino acids (IL-1β), and amino acid analyses of secreted human monocyte IL-1 showed that this corresponds to the carboxyl end of the sequence of the precursor molecules. IL-1s are therefore synthesized as large precursor molecules (35 kd) which are processed at the cell membrane or extracellularly to give the 15 – 17 kd mature active proteins. Biologically active fragments of IL-1 which have been recovered from body fluids such as plasma and urine may represent further breakdown products of the mature protein (12).

Human IL-1β is 27% homologous with IL-1α at the amino acid level. The homology is significant and is mainly at the biologically active carboxy-terminus. Since both IL-1s bind to the same receptor, and to date few distinctions have been reported between them in terms of biological activity, the homologous regions may be important in receptor binding leading to bioactivity. Thus IL-1α and -β are the products of two independently evolving genes with differing but related protein structures and similar biological activities.

IL-1 molecules do not possess classic signal sequences and there are no appreciable hydrophobic portions of the molecule. The absence of 15 – 17 kd proteins inside the cell suggests that processing of the precursor molecule occurs at the membrane or outside the cell although how this might occur is not known. The precise location of the cleavage point of the precursor to give the active fragment is not clear and may be variable. This may explain the disagreement over the size heterogeneity of biologically active molecules. The molecules do not contain disulphide bonds and have no potential glycosylation sites.

IL-1 receptors have been examined by the binding of radiolabelled IL-1 (13). They are found on many cell types in different densities and have a molecular weight of 60 – 80 kd. Fibroblasts have around 150 – 500 per cell and B and T lymphocytes around 50. Upon stimulation of the latter the numbers rise to a few hundred. IL-1α and IL-1β bind to the same receptor with an affinity of $0.2 - 2 \times 10^{-10} \, M^{-1}$. Further characterization by molecular means is awaited.

2.3 Interleukin 2

T cell growth factor (TCGF), now known as IL-2, is a polypeptide produced by activated T cells which acts on T cells to promote their division and other cells of the immune system such as natural killer (NK) and B cells. Human IL-2 was first purified from the culture supernatants of mitogen or alloantigen-activated T cells and the leukaemic cell line Jurkat, and was shown to have a molecular weight of 19 – 22 kd by gel-permeation and 14 – 16 kd on SDS – PAGE (14). Size and charge heterogeneity was attributed to polymerization. As with other lympho-kines, isolation of a cDNA for IL-2 provided the necessary information to re-interpret some of the early biochemistry. The cDNA consists of a single open reading frame coding for 153 amino acids (15). The first 20 amino acids of the amino-terminal end are hydrophobic and constitute the signal sequence which is cleaved off to give the mature protein which consists of 133 amino acids and a predicted molecular weight of 15 420. The difference between this and the observed molecular weight of the natural product arises from O-glycosylation of the threonine residue at position 3 of the mature molecule. All size and charge heterogeneity is therefore due to post-translational modification.

There is a single copy of the human gene consisting of four exons and three introns on chromosome 4 in the human (16) (*Figure 2.6*). The IL-2 molecule

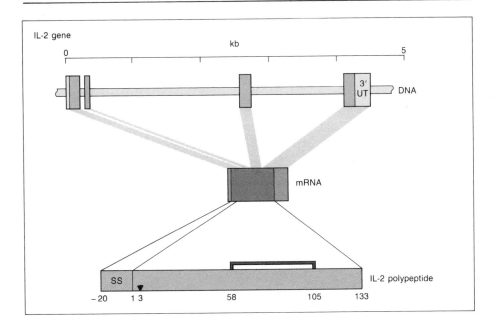

Figure 2.6.   Human IL-2 gene structure. The IL-2 gene consists of four exons which generate a polypeptide of 155 amino acids. The first 20 residues are a signal sequence (SS) cleaved from the mature protein, and there is a disulphide link between cysteine residues at positions 58 and 105. The molecule is O-glycosylated at residue 3.

contains a single disulphide bond between residues 58 and 105 and chemical reduction of the bond or site-directed mutagenesis of these residues leads to loss of biological activity showing that the bond is essential for bioactivity. IL-2 shows no sequence homology with other humoral factors.

IL-2 interacts with cells binding to a receptor of which there are two forms with differing affinities for IL-2. A high affinity receptor K_d of 1×10^{-11} M mediates the physiological response of T cells to IL-2 and about 10% of the receptor sites are of this type. The remaining receptors bind IL-2 with low affinity ($K_d = 1 \times 10^{-8}$ M). Determination of the number of receptor sites per cell by binding studies with either radio-iodinated IL-2 or the anti-receptor antibody (anti-TAC) produced discrepant results, the latter giving higher values than the former (17). The explanation for this became clear once the gene for a 55 kd receptor protein was cloned and transfected into other cells. Transfection of non-lymphoid cells with the human IL-2 receptor gene produced low affinity receptors (18) whereas transfection of T cell lines produced high and low affinity receptors (19). This raises the possibility that lymphocytes can code for a second IL-2 binding protein which could interact with the 55 kd molecule to give rise to high affinity receptors. A second molecule with a molecular weight of 75 kd has been identified on mutant cell lines which binds IL-2 with an affinity intermediate between that of low and high affinity (20). The high affinity receptors are therefore composed of at least two subunits, each of which can independently

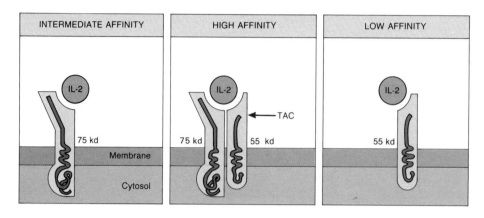

Figure 2.7. IL-2 receptors. The high affinity IL-2 receptor is formed from two non-covalently linked polypeptides, each of which has some affinity for IL-2. The larger (75 kd) peptide contains a longer intracytoplasmic section which is thought to be involved in signalling, while the smaller (55 kd) peptide is recognized by the anti-TAC monoclonal antibody.

bind IL-2 with lower affinity and one of which, the 55 kd protein, may be identified with anti-TAC antibody. Each subunit of the receptor may interact with a different region of the IL-2 molecule and the binding of IL-2 to both proteins produces the biologically active high affinity receptor (21) (*Figure 2.7*).

IL-2 production and high affinity receptor expression are transient events associated with T cell activation [(22) see Chapter 3]. The 75 kd molecule is expressed on resting T cells as well as NK cells. However, the 55 kd molecule is induced on activation and when expressed on the cell surface with the 75 kd molecule gives rise to the high affinity receptor. Binding of IL-2 to the 75 kd molecule may be able to initiate induction of the gene transcribing the 55 kd molecule and hence development of a functional high affinity receptor. This would help to explain the observation that T cells may proliferate in the presence of IL-2 alone. Furthermore, cells such as NK cells which have been reported to be IL-2 receptor negative by staining with anti-TAC, but which proliferate in response to IL-2, may do so by this mechanism (21,23).

2.4 Interleukin 3 and the colony stimulating factors

2.4.1 Interleukin 3

Interleukin 3 (IL-3) is a haematopoietic growth factor which supports the growth and differentiation of pluripotent stem cells leading to different blood cell types, particularly myeloid cells (24). IL-3 is now known to be identical with previously named factors (*Table 1.1*). It induces the expression of 20α-steroid dehydrogenase in splenic lymphocytes, induces Thy-1 expression on lymphocytes, supports the growth of IL-3-dependent cell lines and maintains colony forming units in culture (25). IL-3 is produced in the culture supernatants of activated T helper lymphocytes and is also constitutively produced by some cell lines like the murine

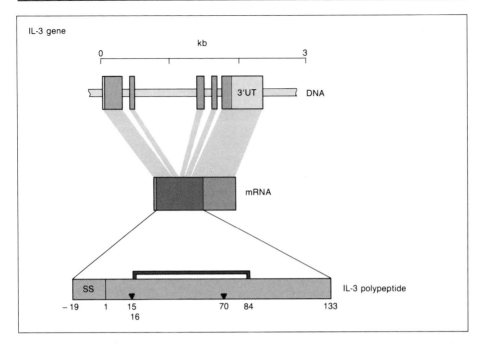

Figure 2.8. IL-3 gene. The structures of mouse and human genes are very similar, each containing five exons. Shading is as in *Figure 2.5*. The polypeptide contains a signal sequence (SS), disulphide bond between residues 16 and 84 and *N*-glycosylation sites at residues 15 and 70.

WEH1-3B cell line. The constitutive production of IL-3 by this line is associated with the insertion of a retrovirus close to the 5' end of the IL-3 gene (see Chapter 5).

The cDNA clones encoding murine IL-3 were described in 1984 (26) whilst the corresponding human sequence remained unknown until 1986, when a cDNA clone encoding IL-3 was isolated from a gibbon T cell line (27). This gibbon cDNA was used as a hybridization probe for the human gene. Since the murine probe had proved ineffective in annealing to human DNA it was concluded that homology between human and mouse molecules would be poor, suggesting that the genes had diverged early in evolution. There is indeed only 29% homology between the human and mouse proteins.

Mouse IL-3 consists of 166 amino acids with a signal sequence of approximately 27 amino acids and four cysteine residues (26). Natural IL-3 has a molecular weight of 28 kd because of glycosylation at the four potential sites. The human gene for IL-3 is located on chromosome 5 and consists of 152 amino acids with a signal sequence of 19 amino acids (27). It has two potential glycosylation sites and two cysteine residues involved in one disulphide bond (*Figure 2.8*). The exon/intron arrangement is similar to the mouse (five exons, four introns).

The IL-3 receptor has not been characterized but binding studies of IL-3 in rats and mice suggest that the receptor systems have evolved separately as have the genes. Thus, there is no interspecies cross-reactivity in binding or function.

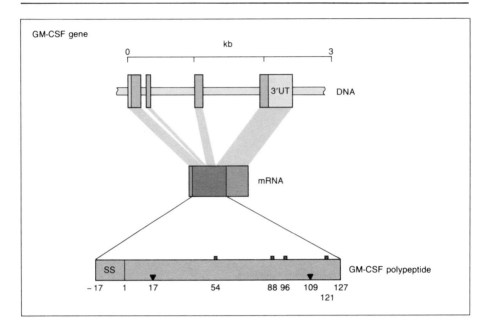

Figure 2.9. Human GM-CSF. The GM-CSF gene contains four exons encoding 144 amino acids, of which 17 are a signal sequence (SS). It is known that a disulphide bond(s) is essential for activity and the positions of the Cys residues are indicated, as well as the two *N*-glycosylation sites at residues 17 and 109.

2.4.2 Granulocyte-macrophage-colony stimulating factor

Granulocyte-macrophage-colony stimulating factor (GM-CSF) stimulates the formation of granulocyte, macrophage, mixed granulocyte-macrophage and, at higher concentrations, eosinophil colonies from pluripotent haematopoietic stem cells. It is produced by activated lymphocytes and a number of cell lines. Medium in which the HTLV-I-infected T lymphoblast Mo cell line had been grown has been used to purify the human protein to homogeneity (28). The resulting glycoprotein had a molecular weight of 22 kd. On SDS – PAGE the natural protein migrates with an apparent molecular mass of between 14 and 35 kd, the range being due to variable glycosylation. A cDNA library prepared from the Mo cell line RNA in an expression vector which transfects monkey COS cells was screened for the transient expression of the lymphokine (29). Human GM-CSF has a single open reading frame coding for a 144 amino acid precursor protein. Seventeen amino acids are cleaved from the amino-terminal end to give a mature 127 amino acid protein. The protein is coded for by a single gene on chromosome 5 in humans and 11 in the mouse (*Figure 2.9*). There are two potential *N*-glycosylation sites and four cysteine residues. Disulphide bonding between the cysteine residues is important for biological activity since reduction leads to loss of biological activity. Mouse and human GM-CSF have 70% nucleotide homology, 60% amino acid homology, are similarly organized and code for homologous proteins, but there is no cross-species reactivity.

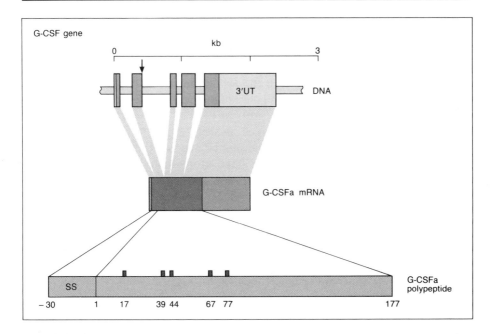

Figure 2.10. Human G-CSF. The gene structure and polypeptide are indicated diagrammatically as in previous figures. Alternative splicing of the primary transcript within the second exon at the point indicated results in mRNA encoding a form of G-CSF with three fewer amino acids. The *O*-glycosylation sites are not shown in this diagram.

Binding studies have been used to study GM-CSF receptors and have shown that all cells in the neutrophil, eosinophil and monocyte series bind GM-CSF. Lymphoid and erythroid cells are negative. There are a small number of receptors per cell (up to a few hundred) and for each cell lineage the numbers decrease with maturation. The receptor has been reported as having a molecular weight of between 51 and 130 kd (30).

2.4.3 Granulocyte-colony stimulating factor

Granulocyte-colony stimulating factor (G-CSF) preferentially stimulates the formation of granulocytic colonies and, at a high concentration, granulocyte-macrophage colonies from pluripotent haematopoietic stem cells. It is made by activated monocytes and macrophages as well as a number of cell lines.

Human G-CSF was purified to homogeneity from the culture supernatants of the human bladder carcinoma cell line 5637 (31). The protein has a molecular weight of 19.6 kd and a pI of 5.5. The full-length cDNA codes for a protein of 207 amino acids of which the first 30 (which are hydrophobic) have been determined as the probable signal sequence (32) (*Figure 2.10*). *O*-Glycanase treatment of the natural protein reduces the molecular weight from 19.6 to 18.8 kd, suggesting that there are potential *O*-glycosylation sites through serine or threonine. There are five cysteine residues which can form intra- and intermolecular disulphide bonds. G-CSF from man and mouse share homology at the amino acid level and exert cross-species biological activity.

The G-CSF gene can code for two polypeptides which have similar biological activity. These arise because two mRNAs may be generated by alternative use of two 5′ splice donor sequences in the second intron of the gene (33).

Binding studies have shown that all cells of the neutrophilic granulocyte series bind G-CSF. No binding has been seen to cells of the erythroid, lymphoid, eosinophilic or megakaryocytic lineages. A small amount of binding has been recorded by normal pro-monocytes and monocytes and there are a few myelo-monocytic/monocytic cell lines which have more than 1000 receptors. The number of receptors per normal cell increases with maturation to a few hundred per cell. The receptor is a single chain of M_r 150 kd (30).

2.4.4 Macrophage-colony stimulating factor

Macrophage-colony stimulating factor (M-CSF) stimulates the formation of macrophage colonies from pluripotent haematopoietic stem cells. It is produced by fibroblast cell lines and is found in urine.

Human M-CSF, purified to homogeneity from urine, showed a molecular weight of 70–90 kd comprised of two identical subunits (35–45 kd). Dissociated subunits are not biologically active. The M-CSF gene consists of many exons covering 22 kb of DNA on chromosome 5 in the human (*Figure 2.11*). At least two differentially spliced mRNAs are expressed by this gene (34,35). The RNAs code for protein precursors which are processed to give mature M-CSF proteins. Thus, a 61 kd protein is coded for by an mRNA which contains a sequence for 298 amino acids inserted within the coding sequence of the smaller (26 kd) precursor. The two precursors share a common amino-terminal sequence follow-ing removal of a 32 amino acid signal sequence, both are processed at the carboxy-terminal, are glycosylated and then associate in homodimers to give the active CSF. At least two natural forms of mature M-CSF exist, a 70–90 kd glycoprotein of dimers of 35–45 kd subunits, comprising approximately 223 amino acids, and a 40–50 kd glycoprotein of dimers of 20–25 kd subunits of around 145 amino acids.

Normal monocytes, macrophages and cell lines display a single class of high affinity receptor with a molecular weight of 165 kd. Receptors are not found on erythroid, lymphoid, eosinophilic or megakaryocytic cells although there may be small numbers on neutrophils. The receptor is a glycoprotein with intra-cytoplasmic tyrosine kinase activity and is identical or closely related to the c-*fms* proto-oncogene also located on chromosome 5. There are an estimated 3–15 000 receptors on macrophages, the number increasing with maturity. Human M-CSF binds to the murine M-CSF receptor but not vice versa (30).

2.5 Interleukin 4

Interleukin 4 (IL-4) causes activation, proliferation and differentiation of B cells. It is also a growth factor for T cells and mast cells and exerts other effects on

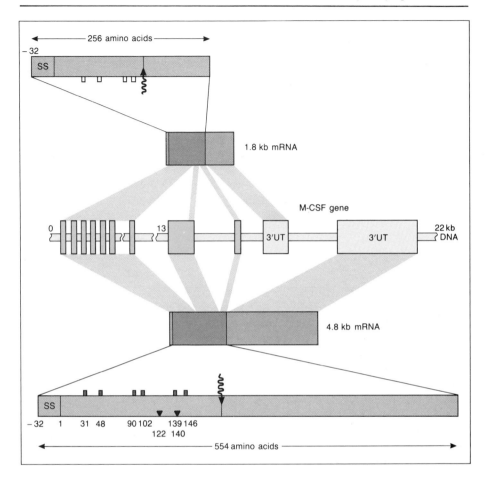

Figure 2.11.   Human M-CSF. The M-CSF gene (centre) extends over 22 kb. There are a large number of small exons in the first 13 kb, and the gene can generate two different mRNAs (1.8 kb and 4.8 kb) and polypeptides by differential splicing of the primary RNA transcript as indicated. The different translation products are cleaved at different positions shown by the wavy arrow, to yield the final polypeptides of approximately 145 or 223 residues. Cysteine residues are at positions 31, 48, 90, 102, 139 and 146: N-glycosylation sites are at 122 and 140.

granulocyte, megakaryocyte and erythrocyte precursors and macrophages (36 – 39). Mouse IL-4 was originally purified from the supernatants of the thymoma line EL-4 which had been stimulated with PMA. It is also generated by activated T cell lines. Natural IL-4 has a molecular weight of 20 kd. The cDNA clones for mouse IL-4 code for 140 amino acids of which the first 20 constitute the signal sequence. The mature 120 amino acid protein has three potential N-glycosylation sites and six cysteines which are all involved in intrachain disulphide bonds (37,38).

A human clone was isolated from a human cDNA library prepared from a

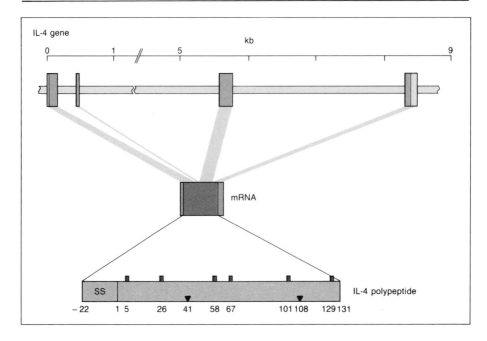

Figure 2.12.   Human IL-4. The human IL-4 gene consists of four exons extending over 9 kb of DNA. The IL-4 polypeptide consists of 153 amino acid residues from which a signal sequence is cleaved. There are two *N*-glycosylation sites and six cysteine residues.

concanavalin-A-stimulated human T cell line (39). This codes for a protein of 153 amino acids of which the first 22 constitute a signal sequence. It has two potential *N*-glycosylation sites and six cysteine residues (*Figure 2.12*). It shares 50% homology with the mouse IL-4. Human IL-4 is inactive on mouse cells as is mouse on human. IL-4 shows no homology with other lymphokines. The human IL-4 gene is located on chromosome 5 and consists of four exons spanning 9 kb.

Receptors for IL-4 are found on a number of cells (B and T cells, macrophages, mast cells and myeloid cells) and have a molecular weight of around 60 kd. There are approximately 300 receptors on B and T cells but more on B cell lymphomas and T cell lines. Receptor numbers increase on cell activation.

2.6 Interleukin 5

Interleukin 5 (IL-5) causes B-cell activation, growth and differentiation factor as well as eosinophil differentiation (40). Natural murine IL-5 purified as a protein of molecular weight 42 – 66 kd, which reduced to 40 kd on SDS – PAGE. The murine cDNA codes for 133 amino acids with a signal sequence of 18 amino acids (H.D.Campbell *et al.*, in preparation). The human cDNA codes for a 134 amino acid protein of which the first 19 are the predicted signal sequence (41). It has two potential glycosylation sites and two cysteine residues. The human gene consists of four exons spanning 3.2 kb on chromosome 5q31 (*Figure 2.13*).

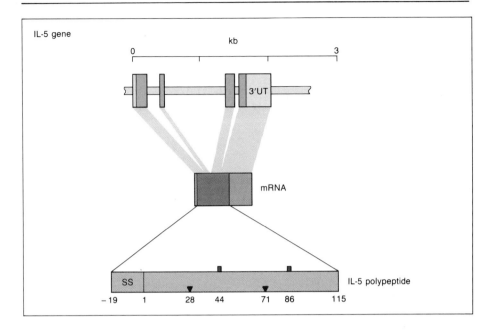

Figure 2.13. Human IL-5. The IL-5 gene consists of four exons which generate a polypeptide of 134 amino acid residues. A 19-residue signal sequence is present (SS) and there are two cysteine residues. There are N-glycosylation sites at residues 28 and 71.

There is 67% homology at the amino acid level between mouse and human IL-5. It is important to note that whilst murine recombinant IL-5 causes both B cell growth and eosinophil differentiation, the human recombinant homologue only causes eosinophil differentiation (40, see Chapter 4).

2.7 Interleukin 6

Interleukin 6 (IL-6), made by activated T cells or T cell lines, has the capacity to stimulate human B cells, is a growth factor for hybridomas and plasmacytomas, and has weak antiviral activity (hence, it has been named IFN-β_2). The purified protein has a molecular weight of 22–29 kd (42). The cDNA codes for a protein of 212 amino acids of which the first 28 form a hydrophobic signal sequence. The molecule has two potential glycosylation sites, and four cysteine residues which may be involved in disulphide bonds (43).

2.8 Cytotoxins

Lymphotoxin (LT) is a lymphokine released by activated T lymphocytes which is cytotoxic and cytostatic for some tumour cell lines *in vitro* and causes haemorrhagic necrosis of certain tumours *in vivo*. It is also released by certain lymphocyte cell lines. It is closely related to tumour necrosis factor (TNF) which was first described as a factor in the serum of mice injected with Bacillus Calmette-Guerin and subsequently challenged with endotoxin, which caused the necrosis

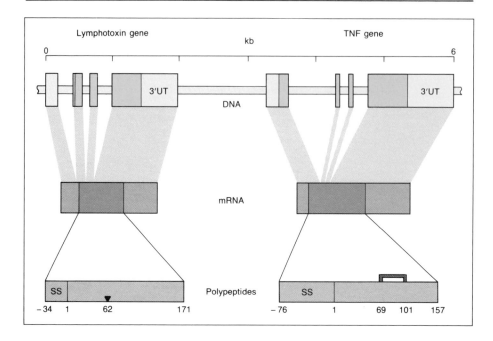

Figure 2.14. The human LT and TNF-α genes lie within the MHC. The gene and protein structures are similar, although TNF has a disulphide bond and LT has an *N*-glycosylation site at residue 62. The genes are closely linked but independently controlled.

of some tumours *in vivo* and was cytostatic for transformed cell lines *in vitro*. TNF is now known to be made by activated macrophages and other cells (*Table 1.3*) and to have wide ranging inflammatory effects. The shared structural and biological properties of these proteins has led to them being renamed TNF-α (TNF) and TNF-β (LT). However, to avoid confusion the terms TNF and LT will be used here.

Natural LT has a molecular weight of 60 – 70 kd and on SDS – PAGE the major species has a molecular weight of 25 kd and 5% of the material a molecular weight of 20 kd (44). The LT gene codes for a protein of 205 amino acids of which the first 34 residues have characteristics of a signal sequence. The mature protein consists of 171 amino acids (18 660 mol. wt). The difference in molecular weights between the cDNA encoded and natural proteins arises from *N*-glycosylation of the mature protein. The 25 kd species has 23 extra amino acids when compared with the 20 kd species (45,46).

TNF was purified from cultured supernatant of stimulated human pro-myelocytic leukaemic line, HL-60, to give a protein of molecular weight of 17 kd and a pI of 5.3 (47). This cell line was used to isolate mRNA for cloning. The isolated cDNA predicted a single open reading frame of 233 amino acids. Since the amino-terminus of the natural product had been determined it could be shown that the first 76 residues were a pre-sequence and that the mature protein had a molecular weight of 17 356, close to that predicted by conventional biochemical

Table 2.1. Human lymphokine structure

Lymphokine	Native protein		cDNA clone		Genomic clone	
	Source	Mol. wt	Amino acids cDNA	Amino acids (mature protein)	Intron/exon	Location (chromosome)
IFN-γ	Peripheral blood leukocytes	40–50 kd (dimers of 20 and 24 kd forms)	166	143	3/4	12(q241)
IL-1α	Monocytes	15–17 kd	271	~154		
IL-1β	Monocytes	15–17 kd	269	~153		
IL-2	T cells	14–16 kd	153	133	3/4	4q
IL-3			152	133	4/5	5
IL-4	T cells	20 kd	153	131	3/4	5
IL-5			134	115	3/4	5
IL-6	Fibroblasts	22–29 kd	212	184	4/5	
GM-CSF	Mo cell line	22 kd	144	127	3/4	5
G-CSF	Human bladder carcinoma	19 kd	207	177	4/5	
M-CSF	Human urine	47–76 kd	224	223 or 145	many	5
TNF	HL-60 cell line	17 kd	233	157	3/4	6
LT	1788 cell line	25 kd	205	171	3/4	6

separation and in agreement with the fact that there were no potential glycosylation sites (48). It is of interest that IL-1α and -β which are also produced by macrophages also have long pre-sequences, and these may be important in macrophage secretory processes. The fact that the pre-sequences in human and mouse TNF are 86% homologous at the amino acid level suggests they may be biologically important.

A significant homology of 35% is found between LT and TNF at the mature protein level, with numerous conservative amino acid changes on alignment. Two regions (48 – 64 and 119 – 133, TNF numbering) are the most conserved in TNF and LT suggesting they may have functional importance. The secreted forms have relatively hydrophilic amino-terminals and significantly hydrophobic carboxy-terminals. It is noteworthy that only LT is glycosylated and that only TNF has cysteine residues which form intrachain disulphide bonds. Both genes contain four exons and three introns and are encoded on human chromosome 6 in the MHC with only 1200 bp separating the two genes (49) (*Figure 2.14*). The two genes are independently regulated and share little homology in their promoter regions. The close linkage of these genes which are preferentially expressed in different cells on the same chromosome is of particular interest.

TNF and LT bind to the same high affinity receptor on various cell lines which all have around 2000 – 3000 binding sites per cell. The amount bound to susceptible cells is proportional to the amount of cell lysis although it is not known whether cell lysis requires internalization of the molecule or simply binding to the receptor. Not all cells that bind proteins are susceptible to their toxic effects, adding a further dimension to the ligand – receptor interaction and the mechanism of killing.

The properties of these lymphokines, and the characteristics of their mRNAs, as determined from cDNA sequences are summarized in *Table 2.1*.

3. Further reading

Gillis,S. (ed.) (1987) *Recombinant Lymphokines and their Receptors.* Marcel Dekker, USA.
Glover,D.M. (ed.) (1985) *DNA Cloning—A Practical Approach,* Vols I – III. IRL Press, Oxford.
Webb,D.R. and Goeddel,D.V. (eds) (1987) *Molecular Cloning and Analysis of Lymphokines.* Academic Press, New York.

4. References

1. Trinchieri,G. and Perussia,B. (1985) *Immunol. Today, 6*, 131.
2. Yip,Y.K., Barrowclough,B.S., Urban,C. and Vilcek,J. (1982) *Science, **215***, 411.
3. Naylor,S.L., Sakaguchi,A.Y., Shows,T.B., Law,M.L., Goeddel,D.V. and Gray,P.W. (1983) *J. Exp. Med., **157***, 1020.
4. Naylor,S.L., Gray,P.W. and Lalley,P.A. (1984) *Somatic Cell Mol. Genet., **10***, 531.
5. Gray,P.W. and Goeddel,D.V. (1982) *Nature, **298***, 859.
6. Celada,A., Gray,P.W., Rinderknecht,E. and Schreiber,R.D. (1984) *J. Exp. Med., **160***, 55.
7. Friedman,R. and Stark,G. (1985) *Nature, **314***, 637.

8. Windle,J.J., Shin,H.S. and Morrow,J.F. (1984) *J. Immunol., 132*, 1317.
9. Kimball,E.S., Pickeral,S.F., Oppenheim,J.J. and Rossio,J.L. (1984) *J. Immunol., 133*, 256.
10. Dinarello,C.A., Clowes,G.H.A., Gordon,A.H., Saravis,C.A. and Wolff,S.M. (1984) *J. Immunol., 133*, 1332.
11. Lomedico,P.T., Gubler,U. and Mizel,S.B. (1987) *Lymphokines, 13*, 139.
12. Auron,P.E., Warner,S.J.C., Webb,A.C., Cannon,J.G., Bernheim,H.A., McAdam,K., Rosenwasser,L.J., Lopreste,G., Mucci,S.F. and Dinarello,C.A. (1987) *J. Immunol., 138*, 1447.
13. Dower,S.K. and Urdal,D.L. (1987) *Immunol. Today, 8*, 46.
14. Robb,R.J. and Smith,K.A. (1981) *Mol. Immunol., 18*, 1087.
15. Matsui,H., Fujita,T., Nishi-Takaoka,C., Hamuro,J. and Taniguchi,T. (1985) *Lymphokines, 12*, 1.
16. Holbrook,N.J., Smith,K.A., Fornace,A.J., Comeau,C., Wiskoci,R.L. and Crabtree, G.R. (1984) *Proc. Natl. Acad. Sci. USA, 81*, 1634.
17. Robb,R.J., Greene,W.C. and Rusk,C.M. (1984) *J. Exp. Med., 160*, 1126.
18. Greene,W.C., Robb,R.J., Svetlik,P.B., Rusk,C.M., Depper,J.M. and Leonard,W.J. (1985) *J. Exp. Med., 162*, 363.
19. Kondo,S., Shimizu,A., Maeda,M., Tagaya,Y., Yodoi,J. and Honjo,T. (1986) *Nature, 320*, 75.
20. Teshigawara,K., Wang,H.M., Kato,K. and Smith,K.A. (1987) *J. Exp. Med., 165*, 223.
21. Wang,H.M. and Smith,K.A. (1987) *J. Exp. Med., 166*, 1055.
22. Cantrell,D.A. and Smith,K.A. (1983) *J. Exp. Med., 158*, 1895.
23. Siegel,J.P., Sharon,M., Smith,P.L. and Leonard,W.J. (1987) *Science, 238*, 75.
24. Nicola,N.A. and Vadas,M. (1984) *Immunol. Today, 5*, 76.
25. Iscove,N.N. and Roitsch,C. (1985) In Sorg,C. and Schimpl,A. (eds), *Cellular and Molecular Biology of Lymphokines*. Academic Press, New York, p. 397.
26. Fung,M.C., Hapel,A.J., Ymer,S., Cohen,D.R., Johnson,R.M., Campbell,H.D. and Young,I.G. (1984) *Nature, 307*, 233.
27. Yang,Y.-C., Ciarletta,A.B., Temple,P.A., Chung,M.P., Kovacic,S., Witek-Giannotti, J.S., Leary,A.C., Kriz,R., Donahue,R.E., Wong,G.G. and Clark,S.C. (1986) *Cell, 47*, 3.
28. Gasson,J.C., Weisbart,R.H., Kaufman,S.E., Clark,S.C., Hewick,R.M., Wong,G.G. and Golde,D.W. (1984) *Science, 226*, 1339.
29. Wong,G.G., Witek,J.S., Temple,P.A., Wilkens,K.M., Leary,A.C., Luxenberg,D.P., Jones,S.S., Brown,E.L., Kay,R.M., Orr,E.C., Shoemaker,C., Golde,D.W., Kaufman, R.J., Hewick,R.M., Wang,E.A. and Clark,S.C. (1985) *Science, 228*, 810.
30. Nicola,N.A. (1987) *Immunol. Today, 8*, 134.
31. Welte,K., Platzer,E., Lu,L., Gabrilove,J.L., Levi,E., Mertelsmann,R. and Moore, M.A.S. (1985) *Proc. Natl. Acad. Sci. USA, 82*, 1526.
32. Souza,L.M., Boone,T.C., Gabrilove,J.L., Lai,P.H., Zsebo,K.M., Murdock,D.C., Chazin,V.R., Bruszewski,J., Lu,H., Chen,K.K., Barendt,J., Platzer,E., Moore, M.A.S., Mertelsmann,R. and Welte,K. (1986) *Science, 232*, 61.
33. Das,S.K., Stanley,E.R., Guilbert,L.J. and Forman,L.W. (1981) *Blood, 58*, 630.
34. Kawasaki,E.S., Ladner,M.B., Wang,A.M., Van Arsdell,J., Warren,M.K., Coyne, M.Y., Schweickart,V.L., Lee,M.T., Wilson,K.J., Boosman,A., Stanley,E.R., Ralph, P. and Mark,D.F. (1985) *Science, 230*, 291.
35. Wong,G.G., Temple,P.A., Leary,A.C., Witek-Giannotti,J.S., Yang,Y.-C., Ciarletta, A.B., Chung,M., Murtha,P., Kriz,R., Kaufman,R.J., Ferenz,C.R., Sibley,B.S., Turner,K.J., Hewick,R.M., Clark,S.C., Yanai,N., Yokota,H., Yamada,M., Saito,M., Motoyoshi,K. and Takaku,S. (1987) *Science, 235*, 1504.
36. Paul,W.E. and Ohara,J. (1987) *Annu. Rev. Immunol., 5*, 429.

37. Noma,Y., Sideras,T., Naito,T., Bergstedt-Lindquist,S., Azuma,C., Severinson,E., Tanabe,T., Kinashi,T., Matsuda,F., Yaoita,Y. and Honjo,T. (1986) *Nature,* **319**, 640.

38. Lee,F., Yokota,T., Ostsuka,T., Meyerson,P., Villaret,D., Coffman,R., Mosmann,T., Rennick,D., Roehm,N., Smith,C., Zlotnik,A. and Arai,K. (1986) *Proc. Natl. Acad. Sci. USA,* **83**, 2061.

39. Yokota,T., Otsuka,T., Mosmann,T., Banchereau,J., De France,T., Blanchard,D., de Vries,J.E., Lee,F. and Arai,K. (1986) *Proc. Natl. Acad. Sci. USA,* **83**, 5894.

40. Sanderson,C.J., Campbell,H.D. and Young,I.G. (1988) *Immunol. Rev.,* **102**, 29.

41. Campbell,H.D., Tucker,W.Q.J., Hort,Y., Martinson,M.E., Mayo,G., Clutterbuck, E.J., Sanderson,C.J. and Young,I.G. (1987) *Proc. Natl. Acad. Sci. USA,* **84**, 6629.

42. Van Damme,J., Opdenakker,G., Simpson,R.J., Rubira,M.R., Cayphas,S., Vink,A., Billiau,A. and Van Snick,J. (1987) *J. Exp. Med.,* **165**, 914.

43. Hirano,T., Yasukawa,K., Harada,H., Taga,T., Watanabe,Y., Matsuda,T., Kashiwamura,S., Nakajima,K., Koyama,K., Iwamatsu,A., Tsunasawa,S., Sakiyama,F., Matsui,H., Takahara,Y., Taniguchi,T. and Kishimoto,T. (1986) *Nature,* **324**, 73.

44. Aggarwal,B.B., Moffat,B. and Harkins,R.N. (1984) *J. Biol. Chem.,* **259**, 686.

45. Gray,P.W., Aggarwal,B.B., Benton,C.V., Bringman,T.S., Henzel,W.J., Jarrett,J.A., Leung,D.W., Moffat,B., Ng,P., Svedersky,L.P., Palladino,M.A. and Nedwin,G.E. (1984) *Nature,* **312**, 721.

46. Nedwin,G.E., Naylor,S.L., Sakaguchi,A.Y., Smith,D., Jarrett-Nedwin,J., Pennica,D., Goeddel,D.V. and Gray,P.W. (1985) *Nucleic Acids Res.,* **13**, 6361.

47. Aggarwal,B.B., Kohr,W.J., Hass,P.E., Moffat,B., Spencer,S.A., Henzel,W.J., Bringman,T.S., Nedwin,G.E., Goeddel,D.V. and Harkins,R.N. (1985) *J. Biol. Chem.,* **260**, 2345.

48. Pennica,D., Nedwin,G.E., Hayflick,J.S., Seeburg,P.H., Derynck,R., Palladino,M.A., Kohr,W.J., Aggarwal,B.B. and Goeddel,D.V. (1984) *Nature,* **312**, 724.

49. Spies,T., Morton,C.C., Nedospasov,S.A., Fiers,W., Pious,D. and Strominger,J.L. (1986) *Proc. Natl. Acad. Sci. USA,* **83**, 8699.

3

Lymphokines in the activation of T cells, B cells and macrophages

1. Lymphokines in T cell activation

Inactive T cells interact with antigen and lymphokines resulting in their proliferation. The clonal expansion of T cells is dependent on the presence of antigen-presenting accessory cells (APCs) which are usually mononuclear phagocytes or dendritic cells. These cells present antigen to the T cell receptor on cytotoxic (TC) and helper (TH) cells in association with class I and class II major histocompatibility complex (MHC) antigens, respectively (*Figure 3.1*).

It is generally held that CD4$^+$ TH cells respond to the dual signals of antigen in association with MHC class II and interleukin 1 (IL-1) generated by the APCs (1) to both proliferate (2) and release a variety of lymphokines which affect T cells, B cells and other cells involved in immune responses (3). There is evidence from work in mice that there are two subpopulations of CD4$^+$ cells which secrete different lymphokines (4). Thus, upon activation, TH-1 cells have been shown to produce mRNA for IL-2, interferon γ (IFN-γ), lymphotoxin (LT), IL-3 and granulocyte-macrophage-colony stimulating factor (GM-CSF) and TH-2 cells for IL-4, IL-5, IL-3 and GM-CSF (*Table 3.1*) (5). The level of mRNA corresponds to the level of lymphokine secreted. It is not known whether these subpopulations exist *in vivo* or whether there are equivalent subpopulations in man. However, the demonstration that different TH lymphocytes might produce different lymphokines suggests that they may be functionally involved in different immune responses.

The clonal expansion of TH and TC lymphocytes is critically dependent on the generation of IL-2 and IL-2 receptors (IL-2^r) (6). The major source of IL-2 for proliferation of both these cell types is the CD4$^+$ TH cell (2). The control of IL-2 production rests at the transcriptional level and induction of the IL-2 gene requires activation of protein kinase C and a rise in intracellular calcium

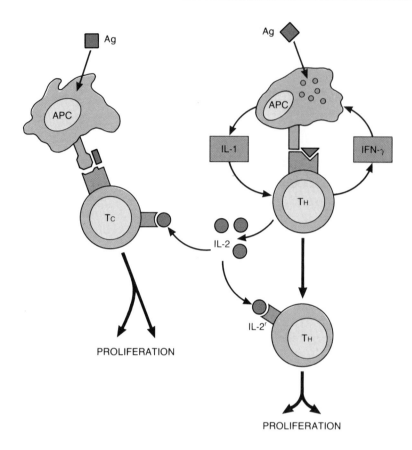

Figure 3.1. T cell activation. T cells are activated by a signal of processed antigen plus MHC on an APC followed by specific lymphokines which drive proliferation and differentiation. IFN-γ from Tн cells can enhance MHC expression and antigen presentation on the APCs. IL-2 is required for proliferation of both Tн and Tc cells. IL-1 from macrophages can potentiate activation by increasing IL-2 receptor expression. (Ag = antigen.)

Table 3.1. Different murine Tн cells make different lymphokines

		IFN-γ	LT	IL-2	IL-3	IL-4	IL-5	GM-CSF
Tн-1	CD4+	+	+	+	+			+
Tн-2	CD4+				+	+	+	+

(7,8). Following stimulation there is transcriptional activation of the IL-2 gene over 24–48 h. The level of secreted IL-2 corresponds well with the level of intracellular IL-2 mRNA, and this, in turn, depends on the level of transcription

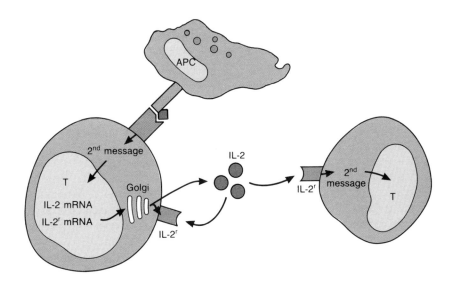

Figure 3.2.   Cellular events in T cell activation. T cell priming induces both IL-2 and IL-2^r expression. The newly synthesized molecules are first detected in the Golgi apparatus. IL-2 acts on other T cells and in an autocrine fashion.

and stabilization of message within the cells (9). IL-2 released from the cells binds to the IL-2^r to lead to clonal proliferation (*Figure 3.2*).

T cells respond to IL-2 via binding to the high affinity IL-2^r made up of the non-covalently linked 75 kd and 55 kd chains (10) (see Chapter 2). The 75 kd chains provide the signal for cell growth whilst binding is facilitated by the 55 kd protein. The 55 kd IL-2^r mRNA production and protein expression, like IL-2 itself, is also induced following binding of antigen – MHC to the CD3 – T cell receptor complex, or of mitogens to their appropriate ligands (11) and is dependent on protein kinase C stimulation (12). Following binding of IL-2 to its receptor, the complex is internalized and therefore removed from the cell surface, and degraded in lysosomes. The half-life at 37°C for this event is 20 – 30 min (12,13).

The time course of receptor expression parallels proliferation with both induction and down-regulation playing an important role in controlling the response (14,15). The number of T cells which become committed to division by passage of the cells from G_1 to S phase of the cell cycle is dictated by the number of occupied IL-2^r on the cell and thus the concentration of IL-2. The proliferative response to IL-2 thus depends on the occupancy of a critical number of high affinity IL-2^r and this will depend on the number of receptors, the IL-2 concentration and the length of the IL-2 – receptor interaction. Removal of the T cell stimulant results in down-regulation of receptor expression and decline in proliferation, thus limiting the response.

The role of IL-1 in the activation of TH cells is not totally clear (16). IL-1 gene transcription and IL-1 production are initiated by a number of substances interacting with macrophages/monocytes. T cells have receptors for IL-1 (17) and exposure of T cells to IL-1 results in transcription of a number of genes including IL-2, IL-3 and the IL-2^r (3). Whether the IL-1^r interacts with soluble IL-1 released from monocytes or a membrane-associated form has not been determined (18). Whilst it was originally believed that IL-1 produced by APCs was an essential component of T cell responses, the current evidence suggests that it has an enhancing rather than obligatory role (16).

It has been shown that IL-1 added to T cells rigorously depleted of accessory cells cannot restore the responses to antigen, and that anti-IL-1 antibodies have no effect on T cell proliferation by antigen. Furthermore, dendritic cells which are potent APCs do not produce IL-1, although their antigen-presenting capacity is enhanced by its presence (19). However, IL-1 does enhance T cell proliferation in the presence of low numbers of APCs and IL-1 results in thymocyte proliferation as seen in the bioassay based on this phenomenon. It may be that in the presence of adequate numbers of accessory cells the adhesion between them and T cells via the antigen – class II MHC interactions may be sufficient. However, in the presence of low numbers of accessory cells, IL-1 is required to optimize the cellular contacts (16,19).

Antigen-driven TH cell proliferation is therefore dependent on the lymphokines IL-1 and IL-2 interacting with appropriate receptors. At the same time other lymphokines are generated (see above) which can themselves feed back onto T cells or macrophages to further amplify the response. Thus, IFN-γ causes increased class II MHC expression on macrophages as well as their increased IL-1 production in the presence of lipopolysaccharide (LPS) (20). Both increased class II MHC expression and IL-1 production serve to activate T cells further. IL-4, also released by antigen-activated T cells, is a T cell growth factor, albeit less potent than IL-2 (21), and would therefore, by its interaction with those T cells bearing an IL-4 receptor, cause proliferation. Tumour necrosis factor (TNF) receptors are induced on activated T cells and, in the presence of TNF, T cells are stimulated to express more receptors for IL-2 and IFN-γ. They also show enhanced IL-2-dependent IFN-γ production (22). These examples illustrate the complex interactions of lymphokines produced by T cells with both the same T cells, other T cells, B cells and macrophages.

2. Lymphokines in B cell stimulation

It has been the tradition to consider that the B cell response to antigen occurs in three sequential steps; namely activation, proliferation and differentiation. Lymphokines which influence B cells were originally divided broadly into B cell growth factors and B cell differentiation factors. The availability of recombinant factors has led to reappraisal of the role they play and of this traditional view.

Activation, proliferation and differentiation may be analysed separately in

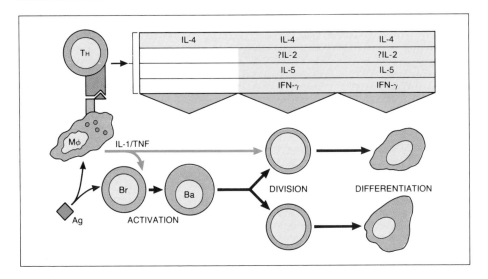

Figure 3.3.   Stages of B cell development in the mouse. Resting B cells (Br) are activated (Ba) by antigen. The processes of activation, division and differentiation are modulated by the macrophage mediators IL-1 and TNF and T-cell-derived lymphokines acting at the points indicated. (Ag = antigen; ? indicates controversial effect.)

laboratory experiments (23 – 25). B cell activation may be examined by looking at the very early events such as changes in phospholipid metabolism, changes in calcium flux, protein phosphorylation and changes in intracellular pH which occur seconds or minutes after ligand receptor binding. Alternatively, they may be measured by later events such as increase in cell size or expression of activation antigens such as MHC class II, the IL-2^r (CD25) or the low affinity Fcε receptor (CD23) occurring some hours after ligand – receptor interaction. B cell proliferation is measured by incorporation of [^{3}H]thymidine into DNA. In practice this is usually undertaken in co-stimulation assays in which B cells are cultured with a growth factor and a low dose of a polyclonal activator such as anti-IgM or LPS. B cell differentiation is measured as the ability of cells to produce antibody. This may be by incubation of B cells with a differentiation factor and an activation signal and examination of antibody production after some days in culture. Alternatively, B cell lines may be incubated with a differentiation factor and assayed for immunoglobulin production. Finally, differentiation may be assessed from the ability of lymphokines to replace TH cells (T cell replacing factors) in antigen-specific antibody responses *in vitro*.

Using these assays, the stages of B cell activation, proliferation and differentiation have been studied in both man and mouse. In both species it is now clear that the same factor may influence all stages, and therefore the idea of separate growth and differentiation factors is not correct. Other factors only seem to operate at one particular stage. Whilst there is broad agreement between man and mouse, there are also some major differences (26) (*Figures 3.3* and *3.4*).

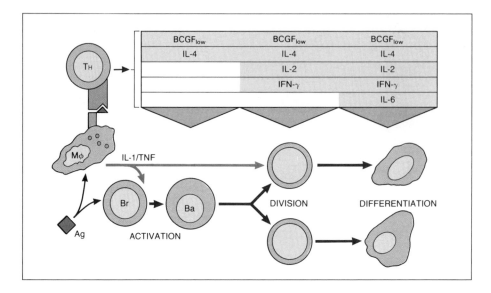

Figure 3.4. Stages of B cell development in human. The processes of activation, division and differentiation are similar to those in mouse, but a different set of lymphokines is active. (Ag = antigen.)

A number of soluble factors have been reported to cause B cell activation, namely IL-4 and a product of T cells referred to as B cell growth factor$_{low}$ (BCGF$_{low}$) because it is only 12 kd in molecular weight. This has recently been cloned in the human (27), although to date most biological information comes from purified natural product (28). These factors may be divided into those which stimulate B cells before they enter the cell cycle (e.g. IL-4, 29) and those which stimulate B cells to enter the cell cycle without completing mitosis (e.g. BCGF$_{low}$, 27). IL-4 causes enhanced expression of a number of cell surface markers including MHC class II and CD23 as well as an increase in cell volume (30,31). IL-1 and TNF have also been described as enhancing factors for B cell activation (32,33).

B cell growth (proliferation) may be stimulated by applying two independent signals to normal B cells, neither of which separately result in growth (25). Such stimuli are a potential growth factor together with either a suboptimal concentration of a polyclonal activator [such as *Staphylococcus aureus* Cowan strain I (SAC) in the human, or anti-IgM in the mouse] or phorbol esters. In these assays IL-2, IL-4, BCGF$_{low}$ as well as TNF and IFN-γ have been reported as causing proliferation (25,26). It is noteworthy that in the mouse IL-5 causes growth, but so far recombinant human IL-5 has failed in many hands to be effective in human assays (34). B cell growth may also be measured using some B cell lines and here again a number of factors have stimulated growth namely IL-2, IL-4 and BCGF$_{low}$ (26). Using the mouse BCL1 line IL-5 has also been reported to cause B cell growth (25).

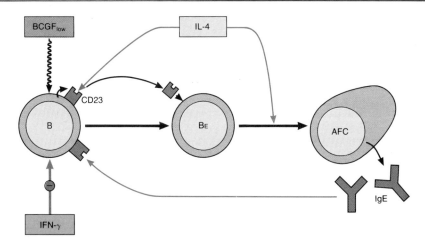

Figure 3.5.   Control of B cell antibody isotype differentiation. BCGF$_{low}$ acts on B cells to cause the cleavage of CD23, the low affinity Fc receptor. This fragment promotes differentiation of IgE-producing B cells (B$_E$) and antibody-forming cells (AFC) under stimulation by IL-4. IgE promotes expression of CD23. IFN-γ inhibits IgE production.

Most of the factors which stimulate B cell growth have been reported as causing B cell differentiation (26). In addition, IL-6 has also been reported as being a differentiation factor without causing growth in the human (35). IL-6 is a potent B cell differentiation factor in inducing IgM and IgG secretion by Epstein – Barr-virus-transformed B cell lines and SAC-activated normal B cells (25). It is as yet unclear whether or how factors may influence production of particular immunoglobulin isotypes. However, it is notable that IL-4 acts as a rather specific differential factor in that it stimulates secretion of IgG$_1$ and IgE (36,37).

A scheme showing the points at which different lymphokines act is shown in *Figures 3.3* and *3.4*. From this it is clear that the same lymphokine may act at different stages of B cell development and that there are many different lymphokines involved. It may be that different factors act on a different, as yet poorly characterized, subpopulation of B cells. Different antigens may use a different repertoire of B cell factors to produce their final effect and may be responsible for different immunoglobulin isotype production. Finally, it is possible that some of the effects seen are only seen in the specialized cultures used *in vitro*. So far, little is known about the micro-environments *in vivo* where lymphokines affecting B cells are generated and exert their action.

The complex regulatory effects of T cells on B cell antibody production are illustrated in *Figure 3.5* and involve BCGF$_{low}$ and the CD23 B cell surface antigen. It has been suggested that BCGF$_{low}$ can cause CD23 to be cleaved to release a soluble product which itself has growth factor activity for B cells (26,38). In the presence of IL-4 these cells would be induced to produce IgE which is the ligand for the low affinity Fcϵ receptor, CD23. Addition of IgE causes up-regulation of CD23 which if cleaved from the surface will stimulate more B

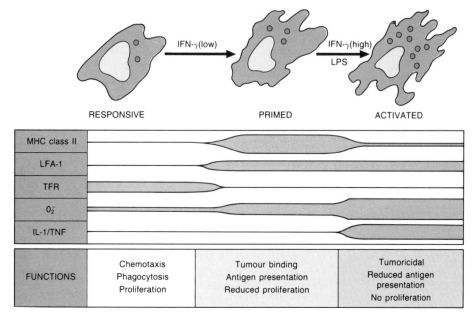

	RESPONSIVE	PRIMED	ACTIVATED

MHC class II			
LFA-1			
TFR			
O_2^-			
IL-1/TNF			

FUNCTIONS	Chemotaxis Phagocytosis Proliferation	Tumour binding Antigen presentation Reduced proliferation	Tumoricidal Reduced antigen presentation No proliferation

Figure 3.6. Activation of macrophages occurs in two stages. The first is driven by low levels of IFN-γ while the second may be driven by LPS or high levels of IFN-γ. Alterations in the levels of expressed molecules and of cell functions are indicated below. TFR is the transferrin receptor.

cell growth thereby completing a positive loop. The up-regulation of CD23 (activation) and production of IgE is inhibited by IFN-γ (39). If IFN-γ is indeed produced by a different helper cell from IL-4 (4) then IgE responses may be controlled at the level of differential stimulation of subsets of T cells.

In the above descriptions B cell growth and differentiation are depicted as controlled by antigen-driven T-cell-derived factors augmented by macrophage-derived factors. There is also evidence that B cells have the capacity to control their own growth in an autocrine fashion. Some B cell lines can produce factors such as IL-4 which stimulate their own growth (40) and there is evidence that normal B cells can also respond to factors they make themselves (26). Such autostimulation must be self-limiting if it is not to lead to uncontrolled growth.

The mechanism by which normal B cell responses are regulated is far from clear. *In vivo* the activation of B cells by T-dependent antigens occurs in lymph nodes in close proximity with the T cells and APCs. In this environment the availability of antigen must limit the response. Presumably, local lymphokine and lymphokine receptor induction occurs according to the availability of the antigen. Little is known about the structure or function of receptors for B cell factors. Studies of receptor and lymphokine expression within the areas of lymph nodes and other lymphoid organs in which antibody is produced should yield important information with respect to the process and its dynamics.

3. Lymphokines in macrophage activation

Monocytes enter tissues to become resident macrophages which are relatively quiescent cells unless they are stimulated, whereupon they change dramatically. This change, known as macrophage activation, results in increased oxygen consumption, increased functional ability (including phagocytosis and tumour cell killing), changes in cell surface markers (including MHC class II) and the secretion of numerous biologically active products including both IL-1 and TNF (20,33). It has long been recognized that T cell activation itself leads to macrophage activation (41), and it is now appreciated that one of the primary T cell products which have macrophage activating activity is IFN-γ (42). A current model (20) (*Figure 3.6*) suggests that macrophages become fully activated in a two stage process in which they respond first to IFN-γ and then to a second signal of which there are many, although most studies have focused on bacterial LPS. It is probable that priming with IFN-γ is not obligatory for release of all lymphokines from macrophages induced by LPS or other substances. However, IFN-γ without doubt amplifies their production. The biologically active products released by activated macrophages include TNF, IL-1 as well as GM-CSF and macrophage-colony stimulating factor. These in turn exert wide ranging effects on homeostasis of the immune system and immunopathology (see Chapter 4).

4. Further reading

Durum,S.K., Schmidt,J.A. and Oppenheim,J.J. (1985) *Annu. Rev. Immunol.,* **3**, 263.
Kishimoto,T. (1985) *Annu. Rev. Immunol.,* **3**, 133.
Möller,G. (ed.) (1986) *Immunol. Rev.,* **92**.
Oppenheim,J.J., Kovacs,E.J., Matsushima,K. and Duram,S.K. (1986) *Immunol. Today,* **7**, 45.
Paul,W.E. and Ohara,J. (1987) *Annu. Rev. Immunol.,* **5**, 429.
Smith,K.A. (1980) *Immunol. Rev.,* **51**, 337.

5. References

1.   Unanue,E.R. and Allen,P.M. (1987) *Science,* **236**, 551.
2.   Palacois,H. (1982) *Immunol. Rev.,* **63**, 73.
3.   Hagiwara,H., Huang,H.J.S., Arai,N., Herzenberg,L.A., Arai,K.I. and Zlotnick,A. (1987) *J. Immunol.,* **138**, 2514.
4.   Mosmann,T.R., Cherwinski,H., Bond,M.W., Giedlin,M.A. and Coffman,R.L. (1986) *J. Immunol.,* **136**, 2348.
5.   Cherwinski,H.M., Schumacher,J.H., Brown,K.D. and Mosmann,T.R. (1987) *J. Exp. Med.,* **166**, 1229.
6.   Watson,J. (1979) *J. Exp. Med.,* **150**, 1510.
7.   Truneh,A., Albert,F., Golstein,P. and Schmitt-Verhulst,A. (1985) *Nature,* **313**, 318.
8.   Weis,A., Wiskocil,R.C. and Stobo,J.D. (1984) *J. Immunol.,* **133**, 123.
9.   Kaempfer,R., Efrat,S. and Marsh,S. (1987) *Lymphokines,* **13**, 59.
10.   Wang,H.M. and Smith,K.A. (1987) *J. Exp. Med.,* **166**, 1055.

11. Leonard,W.J. (1987) *Lymphokines,* **13**, 95.
12. Smith,K.A. and Cantrell,D.A. (1985) *Proc. Natl. Acad. Sci. USA,* **82**, 864.
13. Robb,R.J., Munck,A. and Smith,K.A. (1981) *J. Exp. Med.,* **154**, 1455.
14. Cantrell,D.A. and Smith,K.A. (1983) *J. Exp. Med.,* **158**, 1985.
15. Cantrell,D.A. and Smith,K.A. (1984) *Science,* **224**, 1312.
16. Mizel,S. (1987) *Immunol. Today,* **8**, 330.
17. Dower,S.K. and Urdal,D.L. (1987) *Immunol. Today,* **8**, 46.
18. Kurt-Jones,E.A., Beller,D.I., Mizel,S.B. and Unanue,E.R. (1985) *Proc. Natl. Acad. Sci. USA,* **82**, 1204.
19. Koide,S.L., Inaba,K. and Steinman,R.M. (1987) *J. Exp. Med.,* **165**, 515.
20. Adams,D.O. and Hamilton,T.A. (1987) *Immunol. Rev.,* **97**, 5.
21. Lee,F., Yokota,T., Ostsuka,T., Meyerson,P., Villaret,D., Coffman,R., Mosmann,T., Rennick,D., Roehm,N., Smith,C., Zlotnik,A. and Arai,K. (1986) *Proc. Natl. Acad. Sci. USA,* **83**, 2061.
22. Scheurich,P., Thoma,B., Ucer,U. and Pfizenmaier,K. (1987) *J. Immunol.,* **138**, 1786.
23. Cambier,J.C. and Ransom,J.T. (1987) *Annu. Rev. Immunol.,* **5**, 175.
24. Klaus,G.G.B., Bijsterbosch,M.K., O'Garra,A., Harnett,M.M. and Rigley,K.P. (1987) *Immunol. Rev.,* **99**, 19.
25. Hamblin,A.S. and O'Garra,A. (1987) In Klaus,G.G.B. (ed.), *Lymphocytes—A Practical Approach.* IRL Press, Oxford, p. 209.
26. Gordon,J. and Guy,G.R. (1987) *Immunol. Today,* **8**, 339.
27. Sharma,S., Mehta,S., Morgan,J. and Maizel,A. (1987) *Science,* **235**, 1489.
28. Mehta,S.R., Conrad,D., Sandler,R., Morgan,J., Montagna,R. and Maizel,A.L. (1985) *J. Immunol.,* **135**, 3298.
29. Rabin,E.M., Ohara,J. and Paul,W.E. (1985) *Proc. Natl. Acad. Sci. USA,* **82**, 2935.
30. Defrance,T., Aubry,J.P., Rousset,F., Vanbervliet,B., Bonnefoy,J.Y., Arai,N., Takebe,Y., Yokota,T., Lee,F., Arai,K., de Vries,J. and Banchereau,J. (1987) *J. Exp. Med.,* **165**, 1459.
31. Noelle,R., Krammer,P.H., Ohara,J., Uhr,J.W. and Vitetta,E.S. (1984) *Proc. Natl. Acad. Sci. USA,* **81**, 6149.
32. March,C.J., Mosley,B., Larsen,A., Cerretti,D.P., Braedt,G., Price,V., Gillis,S., Henney,C.S., Kronheim,S.R., Grabstein,K., Conlon,P., Hopp,T.P. and Cosman,D. (1985) *Nature,* **315**, 641.
33. Mahoney,J., Beutler,B., Le Trange,N., Pekala,P. and Cerami,A. (1985) *J. Exp. Med.,* **161**, 984.
34. Sanderson,C.J., Campbell,H.D. and Young,I.G. (1988) *Immunol. Rev.,* **102**, 29.
35. Hirano,T., Yasukawa,K., Harada,H., Taga,T., Watanabe,Y., Matsuda,T., Kashiwamura,S., Nakajima,K., Koyama,K., Iwamatsu,A., Tsunasawa,S., Sakiyama,F., Matsui,H., Takahara,Y., Taniguchi,T. and Kishimoto,T. (1986) *Nature,* **324**, 73.
36. Nomo,Y., Sideras,P., Naito,T., Bergstedt-Lindquist,S., Azuma,C., Severinson,E., Tanabe,T., Kinashi,T., Matsuda,F., Yaoita,Y. and Honjo,T. (1986) *Nature,* **319**, 640.
37. Finkelman,F.D., Katona,I.M., Urban,J.F., Snapper,C.M., Ohara,J. and Paul,W.E. (1986) *Proc. Natl. Acad. Sci. USA,* **83**, 9675.
38. Guy,G.R. and Gordon,J. (1987) *Proc. Natl. Acad. Sci. USA,* **84**, 6239.
39. Coffman,R. and Carty,J. (1985) *J. Immunol.,* **136**, 949.
40. Blazer,B.A., Sutton,L.M. and Strome,M. (1983) *Cancer Res.,* **45**, 4562.
41. Mackaness,G.B. (1968) *Am. Rev. Resp. Dis.,* **97**, 337.
42. Schreiber,R.D., Hicks,L.J., Celada,A., Buchmeier,N.A. and Gray,P.W. (1985) *J. Immunol.,* **134**, 1609.

4

Lymphokines in haematopoiesis, cytotoxicity and inflammation

1. Introduction

Lymphokines are involved in haematopoiesis and the recruitment of cells for host defence. They are important not only for the natural development of white cells during haematopoiesis, but also in the demands for increased numbers of these cells during stress. They are responsible for both the cytotoxic and inflammatory response of a variety of leukocytes to infections and tumours. Produced locally, the effects of the lymphokines are restricted by the tissue structure. However, released systemically, they can affect many tissues of the body in a surprisingly large number of different ways. This chapter deals with the role of lymphokines in the physiology of the immune system with the emphasis on normal host defence mechanisms, whilst the pathology associated with lymphokine release is discussed in Chapter 5.

2. Haematopoietic factors

From *in vitro* data it is believed that a number of lymphokines are involved in the maturation and regulated production of blood cells. This has emerged from studies in which single haematopoietic precursor cells from bone marrow, grown in semi-solid culture systems, develop into discrete colonies having recognizable features of mature cells in the presence of growth factors. These growth factors, termed colony stimulating factors (CSFs), interact at various stages of progenitor cell maturation in a hierarchical fashion (*Figure 4.1*).

Four major CSFs influence haematopoiesis in mouse and, by analogy, in man. Two, interleukin 3 (IL-3) and granulocyte-macrophage-CSF (GM-CSF), have broad specificity, acting on pluripotent stem cells leading to their differentiation, self-renewal and proliferation (1,2), and two others, granulocyte-CSF (G-CSF) and macrophage-CSF (M-CSF), act late in haematopoiesis on cells of particular

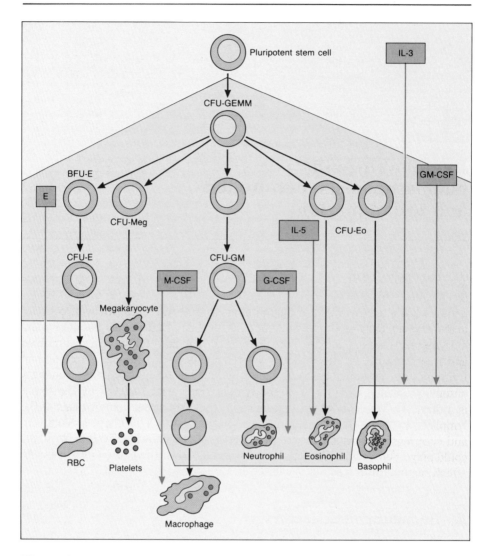

Figure 4.1.   Lymphokine control of haematopoiesis. The lymphokines act in a hierarchical fashion with IL-3 and GM-CSF having wide specificity throughout the differentiation series. M-CSF promotes the macrophage series. G-CSF promotes neutrophils, IL-5 eosinophils, and erythropoietin (E) the differentiation of red blood cells (RBC). (CFU = colony forming unit; BFU = blast forming unit; GEMM = granulocyte, erythrocyte, monocyte, megakaryocyte.)

lineages (3,4). In the presence of IL-3, myeloid progenitors are stimulated to develop into many different cell types such as early erythrocytes, neutrophils, eosinophils, basophils, macrophages and megakaryocytes (1). GM-CSF gives rise mostly to neutrophils, macrophages and eosinophils. In the presence of erythropoietin and GM-CSF or IL-3, erythrocytes and megakaryocytes develop (2). In the presence of G-CSF the colonies are largely made up of neutrophils and their precursors (3), whereas in the presence of M-CSF the colonies consist largely of macrophages (4). These findings suggest a stepwise interaction with IL-3 acting on early progenitors which develop into mature cells of multiple lineages, GM-CSF acting somewhat later to similar effect but G-CSF and M-CSF acting later still to support the growth of late progenitors which are already committed to their particular lineages.

The role these factors play in basal haematopoiesis is unclear. The CSFs have been identified as the products of activated lymphocytes, macrophages and other cells (see *Table 1.3*) (5) and their source in bone marrow, where maturation normally occurs, is not known. Clearly, the stromal cells (endothelium, fibroblasts and perhaps macrophages) would be the most likely sources. Gene probing is needed to identify which, if any, of these stromal cells can produce the lymphokines and under what circumstances, in attempts to determine how basal haematopoiesis is regulated.

In addition to the four CSFs, other factors are also implicated in haematopoiesis. Erythropoietin, which is not a lymphokine, is produced by the kidney and is important for terminal erythrocyte development and regulation of red cell production (6). Haematopoietin 1, now known to be IL-1, primes stem cells to become responsive to the CSFs (7). IL-2, -3, -4, -5 and -6 also exert effects later in the clonal expansion of various cells. Lymphokines which affect antigen-driven T lymphocytes (IL-1 and IL-2) also affect T cell ontogeny within the thymus and may play an important role in clonal expansion and selection there (8). Lymphokines which affect mature B cells may also be important in B cell ontogeny. Both IL-3 and IL-4 are growth factors for mast cells (9,10) and IL-5 is a differentiation factor for eosinophils (11). Tumour necrosis factor (TNF), lymphotoxin (LT) and interferon γ (IFN-γ) have been reported as both stimulating and suppressing haematopoiesis *in vitro*. Thus, many, if not all, lymphokines could play a role in regulating the progression of stem cells towards mature cells which populate the secondary lymphoid organs and peripheral blood.

Lymphokines play important roles in host defence to infection by recruiting cells, activating them and ensuring that there are sufficient cells at times of stress. Activated macrophages and T cells produce CSFs directly and IL-1 and TNF can induce other cells such as fibroblasts and endothelial cells to produce GM-CSF (5). IL-5, which is an eosinophil differentiation factor, may be produced by T cell activation during parasite infections and this might lead to the eosinophilia often accompanying these infections (11). By such mechanisms the immune response to an antigen may result in stimulation of bone marrow cells and their release into the blood stream to undertake important functions during infection.

CSFs released during such immune stimulation can also influence mature cells. Thus GM-CSF activates neutrophils and eosinophils and induces them to undertake antibody-dependent cellular cytotoxicity (ADCC) (12), increase their phagocytosis (13) and become more adherent (14). It enhances macrophage tumoricidal activity *in vitro* (15) and inhibits neutrophil migration (16). G-CSF stimulates mature neutrophils to undertake ADCC, show enhanced phagocytosis and produce superoxide in response to f – Met – Leu – Phe bacterial peptide

(17,18). The large molecular weight version of human M-CSF enhances macrophage cytotoxicity and may be a survival factor for macrophages in tissues. Thus, factors which are believed to be important in normal haematopoiesis are also involved in the immune response to extrinsic antigens both by acting back on the bone marrow and affecting the activity of mature cells.

3. Cytotoxic lymphokines

There are many lymphokines which are important in effecting killing of both infectious agents and tumour cells. This killing may be by the direct action of lymphokines, as is the case for TNF and LT, or by their indirect action by conferring cytotoxicity on cells or augmenting existing cellular cytotoxicity.

LT was originally identified as a factor from mitogen-activated lymphocytes which had anticellular activity for neoplastic cell lines (19). It is cytostatic for certain tumour lines but has little or no normal anticellular activity. TNF was first described by Carswell as an activity present in serum of mice injected with Bacillus Calmette-Guerin and subsequently treated with endotoxin (20). It causes necrosis of certain tumours when injected into tumour-bearing animals and is cytotoxic for a number of transformed cell lines *in vitro*. TNF and LT bind to the same receptor (21) and have the same functional activities. They both cause *in vitro* lysis of actinomycin-D-treated mouse L-929 fibroblasts (22) and necrosis of Meth-A sarcoma tumours *in vivo* (23).

The mechanism by which these substances exert their cytostasis or cytotoxicity is not known (24). TNF and LT interact with the surface receptor and it is probable that internalization is required for cell killing. However, many cells that express receptors are not killed, suggesting that something other than receptor – ligand interactions are required for their cytotoxic action. The effects seen in target cells which may contribute to cell death include fragmentation of DNA, the generation of free radicals, and activation of target cell lysosomal enzymes. However, at the moment the mechanisms of cytotoxicity and the reasons for the variable sensitivity of different target cells to killing by these lymphokines remains unknown.

IFN-γ has antiproliferative activity for numerous cell lines and is therefore a direct cytotoxin (25). IFN-γ also synergizes with LT and TNF in antiproliferative assays *in vitro* (26,27) and, when administered together with LT or TNF, it causes increased antitumour activity. The mechanics of this synergy are not understood but it may be that IFN-γ induces more receptors for the cytotoxins on the tumour cells thereby rendering them more susceptible to cytotoxin action, or it may alter the sensitivity of the target cells to the cytolytic mechanisms outlined above.

The generation of new or augmented cytostatic and cytolytic potential in cells exposed to lymphokines involves many lymphokines and many cell types (*Table 4.1*). Such conferred cytotoxicity may occur on exposure to a lymphokine alone or in combination with a second signal, which may be a second

Table 4.1. Lymphokines inducing or augmenting cytotoxicity

Cell	Effect	Lymphokines
Macrophages	Activation to kill tumour targets	IFN-γ, IL-2
Polymorphs	Augmentation of antibody-dependent cellular cytotoxicity	TNF, LT, GM-CSF, G-CSF
Cytotoxic T cells	Augmentation of MHC-restricted cellular cytotoxicity	IFN-γ
NK cells and LAK cells	Augmentation of MHC-unrestricted killing	IL-2, IFN-γ, IL-1
Eosinophils	Augmentation of killing of antibody-coated tumour cells	IL-5
Eosinophils	Enhanced toxicity for parasites	TNF, GM-CSF, IFN-γ

lymphokine or other stimulant. Amongst these cells, the natural or non-specific major histocompatibility complex (MHC) unrestricted killer cells have received particular attention (28). These natural killers incubated with lymphokines are referred to as lymphokine activated killer (LAK) cells and are capable of greatly augmented killing of tumour cells (29). Their exact phenotype is controversial with reports of CD3, CD8 or CD16 being variably expressed on the effector cells. Similarly their mechanism of killing and the ligands and receptors involved are the subject of debate and controversy.

4. Inflammatory lymphokines

Lymphokines are important in the acute inflammatory response initiated by infection or trauma. The lymphokines predominantly involved are IL-1 and TNF which mediate both local and systemic inflammatory responses which are believed to have survival value. Both IL-1 and TNF are rapidly produced by monocytes and macrophages in response to a number of stimuli such as endotoxin, muramyl dipeptides, lectins, immune complexes and other noxious agents. Bacterial endotoxin is frequently used both *in vivo* and *in vitro* to stimulate their production and study their activity. Once induced, lymphokines can be distributed through the circulation to a large number of sites of activity.

The multiple targets and activities of IL-1 are shown in *Figure 4.2*. This shows that apart from the role IL-1 plays in immunostimulation and haematopoiesis, it also has wide ranging effects on other tissues. Locally (reviewed in 30 and 31), IL-1 induces neutrophil, lymphocyte and monocyte adherence to endothelial cells, stimulates endothelial cell procoagulant activity and plasminogen activator inhibitor synthesis. These effects on endothelial cells are important in limiting the spread of infections allowing leukocytes to pass and be retained at inflammatory sites. Other local changes brought about by IL-1 include enzyme release by osteoclasts, chondrocytes and fibroblasts, and proliferation of epithelial, endothelial, synovial cells and fibroblasts. These may have importance in both

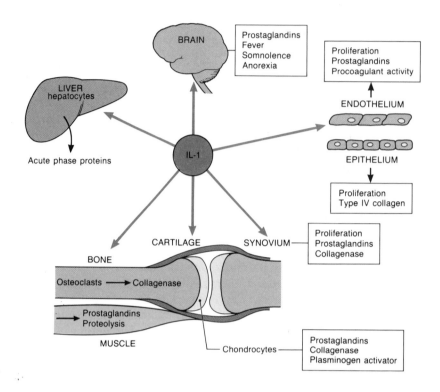

Figure 4.2.   Actions of IL-1. IL-1 has numerous actions on cells other than those of the immune system as indicated.

local inflammation as well as wound healing. IL-1 induces proteolysis of muscle which releases amino acids for synthesis of new proteins. In addition IL-1 produces systemic acute phase changes including neutrophilia, hypozincaemia, hypoferremia and increased synthesis of acute phase proteins by hepatocytes. It also acts on the brain to initiate fever, adrenocorticotrophic hormone release, and sleep. These systemic effects are also considered of value in host defence against infection. For example, T and B cell activity are enhanced by fever (32) and release of acute phase proteins and complement assist in the clearance of micro-organisms by the reticulo-endothelial system (33). Little is known concerning the biochemical effects following IL-1 binding to target cells although it has been reported to depend on calcium channel formation and to lead to activation of arachidonic acid metabolism and the release of prostaglandins, thromboxanes and leukotrienes (30).

  In a similar fashion TNF is also able to produce wide ranging effects on a variety of different cells which may be considered beneficial in the acute response to infection or trauma (reviewed in ref. 34). Curiously, it can both stimulate and suppress mRNA synthesis for different genes. Thus, it suppresses the synthesis of mRNA for lipoprotein lipase in adipocytes but induces the biosynthesis of MHC antigens (35), GM-CSF (5) and IL-1 (36). These latter effects would serve to augment an existing immune response.

It is clear that IL-1 and TNF share many biological activities. This may be because TNF and IL-1 by binding to their respective receptors can produce the same end effects. However, since TNF can induce IL-1 production, many of the effects seen may arise from induction of IL-1 by TNF. LT binds to the same receptor as TNF and has the same spectrum of biological activities. Thus, IL-1, TNF and LT are all powerful inflammatory mediators. The extent to which they each contribute to inflammation will depend on the types of cells activated in a particular response and the ways in which their production is regulated.

5. Further reading

Clark,S.C. and Kamen,R. (1987) *Science,* **236**, 1229.
Kluger,M.J., Oppenheim,J.J. and Powanda,M.C. (eds) (1985) *Progress in Leucocyte Biology,* Vol. 2. Alan R.Liss, New York.
Metcalf,D. (1984) *The Hemopoietic Growth Factors.* Elsevier, Amsterdam.
Metcalf,D. (1986) *Blood,* **67**, 257.
Sieff,C.A. (1987) *J. Clin. Invest.,* **79**, 1549.
Stutman,O. and Lattime,E.C. (1985) *Lymphokines,* **12**, 107.

6. References

1.   Schrader,J.W. (1986) *Annu. Rev. Immunol.,* **4**, 205.
2.   Sieff,C.A., Emerson,S.G., Donahue,R.E., Nathan,D.G. and Wang,E.A. (1985) *Science,* **230**, 1171.
3.   Metcalf,D. and Nicola,N.A. (1983) *J. Cell Physiol.,* **116**, 198.
4.   Stanley,E.R. and Heard,P.M. (1977) *J. Biol. Chem.,* **252**, 4305.
5.   Munker,R., Gasson,V.J., Ogawa,M. and Koeffler,H.P. (1986) *Nature,* **323**, 79.
6.   Fisher,J. (1983) *Proc. Soc. Exp. Biol. Med.,* **173**, 289.
7.   Warren,D. and Moore,M.A.S. (1988) *Proc. Natl. Acad. Sci. USA,* in press.
8.   Hamblin,A.S. and Edgeworth,J. (1988) In Kendall,M. and Ritter,M. (eds), *Thymus Update,* **1**, 135.
9.   Shrader,J., Lewis,S.J., Clark-Lewis,I. and Culvenor,J.G. (1981) *Proc. Natl. Acad. Sci. USA,* **78**, 323.
10. Mosmann,T.R., Bond,M.W., Coffman,R.L., Ohara,J. and Paul,W.E. (1986) *Proc. Natl. Acad. Sci. USA,* **83**, 5654.
11. Sanderson,C.J., Campbell,H.D. and Young,I.G. (1988) *Immunol. Rev.,* **102**, 29.
12. Vadas,M.A., Nicola,N.A. and Metcalf,D. (1983) *J. Immunol.,* **130**, 795.
13. Metcalf,D., Begley,C.G., Johnson,G.R., Nicola,A., Vadas,N.A., Lopez,A.F., Williamson,D.G., Wong,G.G., Clark,S.C. and Wang,E.A. (1986) *Blood,* **67**, 37.
14. Arnaout,M.A., Wang,E.A., Clark,S.C. and Sieff,C.A. (1986) *J. Clin. Invest.,* **78**, 597.
15. Grabstein,K.H., Urdal,D.L., Tushinski,R.J., Mochizuki,D.Y., Price,V.L., Cantrell, M.A., Gillis,S. and Conlon,P.J. (1986) *Science,* **232**, 506.
16. Gasson,J.C., Weisbart,R.H., Kaufman,S.E., Clark,S.C., Hewick,R.M., Wong,G.G. and Golde,D.W. (1984) *Science,* **226**, 1339.
17. Platzer,E., Welte,K., Gabrilove,J.L., Paul,L., Harris,P., Mertelsmann,R. and Moore, M.A.S. (1985) *J. Exp. Med.,* **162**, 1788.

18. Lopez,A.F., Williamson,J., Gamble,J.R., Begley,C.G., Harlan,J.M., Klebanoff,S.J., Waltersdorph,A., Wong,G., Clark,S.C. and Vadas,M.A. (1986) *J. Clin. Invest.*, **78**, 1220.
19. Ruddle,N.H. and Waksman,B.H. (1968) *J. Exp. Med.*, **128**, 1267.
20. Carswell,E.A., Old,L.J., Kassel,R.L., Green,S., Fiore,N. and Williamson,B. (1975) *Proc. Natl. Acad. Sci. USA*, **72**, 3666.
21. Aggarwal,B.B., Eessalu,T.E. and Hass,P.E. (1986) *Nature*, **318**, 665.
22. Mathews,N. (1978) *Br. J. Cancer*, **38**, 310.
23. Gray,P.N., Aggarwal,B.B., Benton,C.V., Bringman,T.S., Henzel,W.J., Jarrett,J.A., Leung,D.W., Moffat,B., Ng,P., Svedersky,L.P., Palladino,M.A. and Nedwin,G.E. (1984) *Nature*, **312**, 721.
24. Ruddle,N.H. (1987) *Immunol. Today*, **8**, 129.
25. Shalaby,M.R., Hamilton,E.B., Benninger,A.H. and Marafino,B.J. (1985) *J. Interferon Res.*, **5**, 339.
26. Stone-Wolff,D.S., Yip,Y.K., Kelker,H.C., Lee,J., Henriksen-De Stefano,D., Rubin, B.Y., Rinderknecht,E., Aggarwal,B.B. and Vilcek,J. (1984) *J. Exp. Med.*, **159**, 828.
27. Lee,S.H., Aggarwal,B.B., Rinderknecht,E., Assisi,F. and Chiu,H. (1984) *J. Immunol.*, **133**, 1083.
28. Hersey,P. and Bolhuis,R. (1987) *Immunol. Today*, **8**, 233.
29. Grimm,E.A., Mazumder,A., Zhang,H.Z. and Rosenberg,S.A. (1982) *J. Exp. Med.*, **155**, 1823.
30. Dinarello,C.A. (1984) *Rev. Infect. Dis.*, **6**, 51.
31. Oppenheim,J.J., Kovacs,E.J., Matsushima,K. and Durum,S.K. (1986) *Immunol. Today*, **7**, 45.
32. Jampel,H.D., Duff,G.W., Gershon,R.K., Atkins,E. and Duram,S.K. (1983) *J. Exp. Med.*, **157**, 1229.
33. Pepys,M.B. and Baltz,M. (1983) *Adv. Immunol.*, **34**, 141.
34. Beutler,B. and Cerami,A. (1986) *Nature*, **320**, 584.
35. Scheurich,P., Kronke,M., Schluter,C., Ucer,U. and Pfizenmaier,K. (1986) *Immunobiology*, **172**, 291.
36. Nawroth,P.P., Bank,D., Handley,J., Cassimeris,J., Chess,L. and Stern,D. (1986) *J. Exp. Med.*, **163**, 1363.

5

Lymphokines in pathology and therapy

1. Lymphokines and pathology

Since lymphokines are important in the regulation of immune responses it follows that their over- or under-production may be involved in the pathology of diseases which directly or indirectly involve the immune system. Furthermore, their potent ability to affect immune cells and their availability in recombinant form provides the rationale and means for their use in immunotherapy. In this chapter the role of lymphokines in disease and therapy are considered.

There are many reports that production of lymphokines or expression of lymphokine receptors is abnormal in different diseases. Such studies have been prompted by the idea that lymphokines, which play an important part in immune responses to extrinsic antigens, might also play a role in immune responses to self-antigens and therefore in the initiation and maintenance of auto-immunity. Since they are involved in acute inflammatory processes they are likely to play a part in chronic inflammatory diseases. Finally, since lymphokines are growth factors for cells of the immune system it might be expected that they may be involved in the disregulated growth of lymphoid and myeloid tumours.

Presentation of auto-antigens may occur when cells which normally do not express class II major histocompatibility complex (MHC) products are induced to do so (1,2). Thyrocytes from patients with auto-immune thyroid disease (3) and beta cells in the 'diabetic' pancreas (4) express class II MHC products, whereas their normal counterparts do not. These cells may bypass the need for normal antigen-presenting cells, and be able to present self-antigens to T cells (5). Activated T cells would then be clonally expanded, become effector T cells and induce auto-antibody production by B cells. These processes would involve the lymphokines important in normal T and B cell activation (Chapter 3). In this model the induction of class II molecules is critical to the development of auto-immunity. Since interferon γ (IFN-γ) is a potent inducer of these molecules, inappropriate production of this lymphokine would lead to class II expression

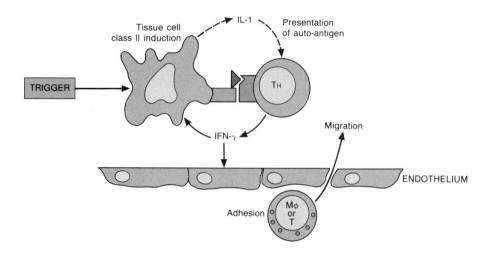

Figure 5.1. Role of IFN-γ in the development of auto-immune reactions. Tissue cells respond to IFN-γ by expressing class II MHC molecules and may present antigen to T cells. Some tissue cells can also produce IL-1 when stimulated, which enhances presentation. IFN-γ also acts on endothelium to enhance macrophage and T cell adhesion and migration.

(*Figure 5.1*). This might occur as a result of viral infections, which have been implicated in the induction of some endocrine auto-immune diseases (6). It has yet to be proven that all these processes can and do occur in the development of an auto-immune disease. However, the model implies that lymphokine production in the wrong place at the wrong time might allow an auto-immune response to emerge with subsequent damaging consequences.

Interleukin 1 (IL-1) and tumour necrosis factor (TNF) are important in acute inflammatory responses manifesting both local and systemic effects. These same activities may be important in the maintenance of chronic inflammation such as occurs in rheumatoid arthritis or multiple sclerosis. Here, it would be envisaged that lymphokines are involved in the generation and amplification of immune responses to unknown extrinsic or self-antigens as described above and dis-regulated production of IL-1 and TNF would mediate tissue damage. Released locally, IL-1 and TNF in humans can activate vascular endothelial cells resulting in increased adhesiveness of leukocytes to them and this may result in diapedesis of inflammatory cells into tissues (7). IL-1 and TNF activate osteoclasts, leading to bone resorption (8), and stimulate chondrocytes, fibroblasts and other cells like synoviocytes to release degradative enzymes which may destroy cartilage (9). These processes are thought important in leukocyte infiltration and erosive damage of, for example, cartilage and bone seen in several chronic inflammatory diseases (10).

The systemic effects of TNF and IL-1 are also important in the inflammatory response accompanying many diseases. In moderation these may be considered beneficial to the host (see Chapter 4). However, uncontrolled production may

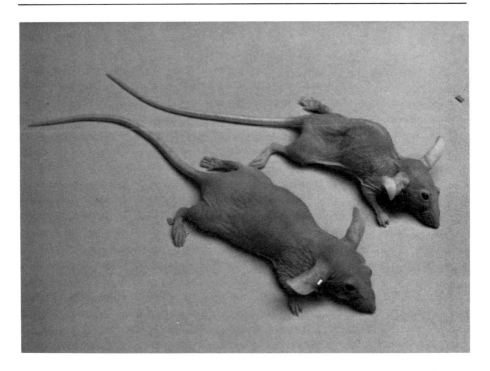

Figure 5.2. Induction of cachexia by TNF. Nude mice which were given CHO tumour cells containing the TNF gene insert are severely wasted (right) by comparison with animals which have the same tumour cells lacking the TNF gene (left). By courtesy of Dr A.Oliff, reprinted by permission of *Cell* (16). Copyright is held by Cell Press.

be damaging. It has been known for a long time that Gram-negative bacterial infection may lead to endotoxic shock characterized by, for example, fever, metabolic acidosis, diarrhoea, hypotension and disseminated intravascular coagulation (11). Bacterial endotoxins are thought to induce these systemic toxic effects by stimulating IL-1 and TNF production. This is supported by the fact that TNF is a major product of macrophages following lipopolysaccharide stimulation (12), amounting to 1–2% of the total secretory protein. After injection of endotoxin into animals, large amounts of TNF appear in the circulation within minutes, peaking 1–2 h later (13). In addition, injection of recombinant TNF into animals produces endotoxic shock (14). This bacterial endotoxin may induce TNF and IL-1 in very large amounts which may lead to the signs and symptoms of shock.

TNF is also called cachectin and is believed to be responsible for the wasting of cachexia associated with cancer and infectious disease. It depresses lipoprotein lipase production and this prevents the uptake of exogenous triglyceride and hydrolysis to free fatty acids and glycerol (15). Rodents implanted with tumours continuously secreting TNF from a transfected human TNF gene become wasted (16) (*Figure 5.2*) and infusion of TNF into animals causes anorexia. Thus, it seems that TNF can produce cachexia, although it is not clear how it is generated in diseases such as cancer. The tumour cells themselves may synthesize TNF, or may stimulate macrophages to do so. In infectious diseases, overwhelming continuous stimulation of macrophages may lead to sufficient generation of TNF to result in cachexia.

It is an attractive hypothesis that lymphokines are involved in the disregulated growth of lymphoid or myeloid cells resulting in leukaemia. Since lymphokines can act in an autocrine fashion, it is conceivable that leukaemic cells may emerge because they both produce and respond to growth factors in an uncontrolled fashion. Alternatively, they may grow because they have lost their regulated control by lymphokines. In leukaemias there is only suggestive evidence for autocrine lymphokine action. For example there is a report that a few acute myeloid leukaemia patients have blast cells which are autocrine for granulocyte-macrophage-colony stimulating factor (GM-CSF) (17). However, in most chronic and acute myeloid leukaemias, GM-CSF is required as a growth factor *in vitro* (18), supporting the view that within the leukaemic cell population there is a selective advantage in expression of the GM-CSF receptor, but that they are not able constitutively to produce GM-CSF. The role of constitutive lymphokine or lymphokine receptor expression in the development of leukaemias therefore remains an unsolved area of future interest.

Lymphokine gene expression can be altered by the nearby insertion of a retrovirus. The WEH1-3B cell line, which produces a myelomonocytic leukaemia in mice, is also a constitutive producer of IL-3 *in vitro*. It has been shown that there is a 5 kb insert 5' to the gene which is a type of endogenous murine retrovirus (19). This insertion leads to up-regulated expression of IL-3. However, the relationship of the IL-3 gene activation to the development of the WEH1-3B leukaemia has yet to be fully defined. Retroviruses can, however, be used to manipulate lymphokine gene expression in experimental animals and these animal models are used to investigate the pathophysiological effect of abnormal lymphokine gene expression. Retroviral expression vectors carrying such lymphokine genes as IL-3 and GM-CSF have been used to study the consequences of continuous gene expression *in vivo* and *in vitro*. Animals into which lymphokine genes have been introduced in this manner have profound changes in circulating leukocytes of the appropriate types which infiltrate tissue and show accompanying pathophysiological changes.

Lymphokine gene expression has been studied in diseases associated with retroviral infection. Most T cells infected with retrovirus HTLV-I, the agent responsible for adult T cell leukaemia, grow *in vitro* without the addition of exogenous IL-2 and constitutively express the IL-2 receptor (20) suggesting that cell growth is dependent on the constitutive production and autocrine action of IL-2. However, the cells not only fail to produce IL-2 constitutively but also cannot be induced to produce IL-2 upon stimulation. In contrast, the IFN-γ gene is both active and inducible suggesting that HTLV-I does not cause generalized repression of gene expression. Rather it seems that cell growth without IL-2 control may be the very reason for their establishment as leukaemic cells.

Acquired immune deficiency syndrome (AIDS) is associated with the retrovirus HIV which infects T cells and leads to their progressive depletion *in vivo*. Since a lack of IL-2 might contribute to this loss, IL-2 gene transcription in HIV-infected cells has been studied (20). However, IL-2 production may be induced to normal levels, and it seems unlikely that lack of inducible IL-2 is important in T cell loss in AIDS. In this retroviral infection the IFN-γ gene is inactive and uninducible. Thus, HIV infection produces a different pattern of gene expression from HTLV-I infection. These examples show that retrovirus insertion can be associated with abnormal expression of lymphokine genes, although the role this plays in development of disease is unclear.

Immunodeficiencies which are directly caused by deletion or complete suppression of lymphokine genes have so far not been described. It is of interest that many of the genes for lymphokines are located on chromosome 5 and that deletions on the long arm of chromosome 5 are frequently observed in patients with myelodysplastic syndrome, acute non-lymphocytic leukaemia secondary to cytotoxic therapy, as well as refractory anaemia characterized as a '5q-syndrome' (21). However, a causal relationship between the deletion part of chromosome 5 and these diseases has not been established. It is interesting to speculate on the effect of deletion of a lymphokine gene on the survival of the individual. Would other genes coding for lymphokines with similar activities compensate for the loss or would the deletion be lethal? The answers to these questions await further studies.

2. Lymphokines and therapy

The interest in lymphokines and therapy arises both from their potential direct use as well as the need to inhibit their production or action. The evidence that they are involved in chronic inflammation (10,22) has resulted in a burgeoning interest in both natural (23) and synthetic inhibitors. However, since the same mediators also regulate normal immune responses, it will be necessary to influence abnormal lymphokine production and action without damaging normal immunoregulatory networks.

Direct therapy with lymphokines has a number of applications. Firstly, lymphokines may be used to reconstitute a failed immune system as in AIDS or to recruit cells needed to overcome temporary immunodeficiency such as that following cytotoxic therapy for treatment of tumours or bone-marrow transplantation. Secondly, lymphokines may also be used to stimulate the host immune response to tumours or overwhelming infections. In all these areas work has begun using recombinant lymphokines in both experimental models in animals and phase I trial studies in humans.

Lymphokines which affect haematopoiesis have considerable clinical potential in acute regeneration after cytotoxic treatment, bone-marrow transplantation, aplastic anaemia, agranulocytosis (idiopathic, toxic or genetic), congenital or acquired neutrophil dysfunction and infections. Injection of lymphokines into

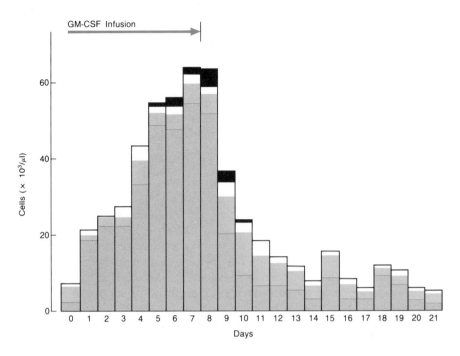

Figure 5.3. Effect of GM-CSF *in vivo*. Normal monkeys were infused with 500 units of human recombinant GM-CSF min^{-1} kg^{-1} for 1 week. Blood counts show a massive increase in neutrophils (grey) and monocytes (white) and pronounced eosinophilia (black) at 4–9 days. Lymphocytes (orange) are also increased, but to a lesser extent. Data of Donahue *et al.*, ref. 24.

both normal and diseased recipients results in the rapid appearance of cells in the blood and tissues. Injection of human recombinant GM-CSF into monkeys (*Macaca fasicularis* or *M.mulata*) gave a dramatic dose-dependent leukocytosis, consisting largely of neutrophils, within 24–48 h (24) (*Figure 5.3*). This was maintained for up to 1 month by continuous infusion without substantial side effects. Upon termination of infusion the leukocyte numbers returned to normal in 3–7 days. Administration of GM-CSF to monkeys rendered pancytopaenic with a Simian type D retrovirus (24) and monkeys recovering from autologous bone-marrow transplantation (25) showed similar effects. In all cases the elevation was transient and dependent on continuous infusion of the GM-CSF. Murine IL-3 stimulated progenitor cell proliferation in both normal and sublethally irradiated mice (26) and human recombinant granulocyte-CSF administered to normal and cyclophosphamide-treated monkeys caused an increase in peripheral neutrophil count (27).

Stimulation of the immune response with lymphokines has application to treatment of infectious diseases and tumours. There is some evidence that IFN-γ may have a beneficial effect in lepromatous leprosy by causing both enhanced

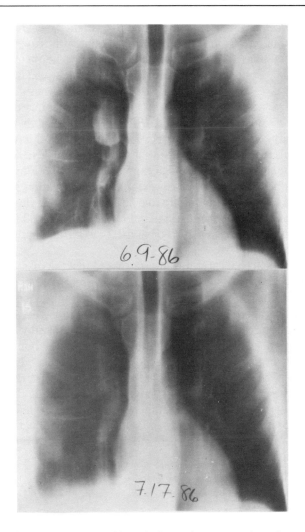

Figure 5.4. Linear tomograms of lung lesions of a metastatic melanoma before (top) and after (bottom) treatment with a combination of IL-2 and LAK cells. The parenchymal and mediastinal lesion on the left of the top figure regressed appreciably after therapy. By courtesy of Dr S.Rosenberg, reprinted by permission of *The New England Journal of Medicine* (31).

immunological responses and some reduction in *Mycobacterium leprae* burden (28). Such studies will be extended when there is further information on the acute toxicity of lymphokines and their effects in animal models of infectious disease.

The application of lymphokines to stimulation of immune responses has mostly been evaluated in patients with tumours. The finding that leukocytes incubated with IL-2 gave rise to a cytotoxic cell capable of killing tumour cells but not normal cells (29) prompted both laboratory investigation and application to human tumour therapy. Lymphokine activated killer (LAK) cells were shown to be effective in mediating regression of lung and liver metastases in various murine models (30). Although IL-2 alone in very large amounts was shown to affect these tumours the injection of IL-2 with LAK cells proved to be the most beneficial.

These laboratory studies have led to clinical trials in which LAK cells are administered with IL-2 in humans (31). In this system patients' leukocytes are incubated *ex vivo* with IL-2 and then these are re-administered with IL-2. Significant regressions were reported (*Figure 5.4*) although there was also significant toxicity associated with the treatment. The IL-2 gave rise to leaky capillaries and this resulted in fluid retention which caused dose-limiting toxicity.

Other potential agents for tumour therapy are IFN-γ, TNF and lymphotoxin (LT). IFN-γ has been shown to cause macrophage activation in clinical trials (32), but like IFN-α it has dose-limiting toxic effects including flu-like symptoms, fatigue, confusion, and gastrointestinal and cardiac toxicity. The potential for low doses of IFN-γ together with TNF and LT as a tumour therapy is under investigation.

A number of clinical trials are thus in progress to test the toxicity and efficacy of lymphokines. These studies have produced a great deal of information on which to base future programmes. The half-life of lymphokines is minutes in the circulation. Therapeutic benefit is only likely to be achieved when the effects lymphokines produce can be achieved by short, intense therapy or maintenance of sufficiently high levels without severe toxicity. Since lymphokines work together, both synergistically and in cascades, it is likely that 'cocktails' of them may produce greater benefit than a single factor. For example, the fact that IL-1 is the same as haematopoietin 1 and renders progenitor cells more responsive to CSFs would suggest that IL-1 administered with CSFs may offer further advantage. The process of determining the safest, and most effective combination of lymphokines will slowly lead to new therapeutic regimes for the treatment of selected diseases.

3. Further reading

Beutler,B. and Cerami,A. (1987) *N. Engl. J. Med.,* **316**, 379.
Billingham,M.E.J. (1987) *Br. Med. Bull.,* **43**, 350.
Bottazzo,G.F., Todd,I., .Mirakian,R., Belfiore,A. and Pujol-Borrell,R. (1986) *Immunol. Rev.,* **94**, 137.
Fauci,A.S., Rosenberg,S.A., Sherwin,S.A., Dinarello,C.A., Longo,D.L. and Lane,H.C. (1987) *Ann. Int. Med.,* **106**, 421.

4. References

1. Hanafusa,T., Pujol-Borrell,R., Chiovato,L., Russell,R.C.G., Doniach,D. and Bottazzo, G.F. (1983) *Lancet,* **ii**, 1111.
2. Bottazzo,G.F., Pujol-Borrell,R., Hanafusa,T. and Feldmann,M. (1983) *Lancet,* **ii**, 1115.
3. Todd,I., Pujol-Borrell,R., Abdul-Karim,B.A.S., Hammond,L.J., Feldmann,M. and Bottazzo,G.F. (1987) *Clin. Exp. Immunol.,* **69**, 532.

4. Pujol-Borrell,R., Todd,I., Doshi,M., Gray,D., Feldmann,M. and Bottazzo,G.F. (1986) *Clin. Exp. Immunol.*, **65**, 128.
5. Londei,M., Bottazzo,G.F. and Feldmann,M. (1985) *Science*, **228**, 85.
6. Todd,I., Pujol-Borrell,R., Hammond,L.J. and Bottazzo,G.F. (1985) *Clin. Exp. Immunol.*, **61**, 265.
7. Granstein,R.D., Margolis,R., Mizel,S.B. and Sauder,D.N. (1986) *J. Clin. Invest.*, **77**, 1020.
8. Gowen,M., Wood,D.D., Ihrie,E.J., McQuire,M.K.B. and Russell,R.G.G. (1983) *Nature*, **306**, 378.
9. Gordon,A.H. and Koj,A. (1985) *Res. Mon. Cell Tiss. Physiol.*, **10**.
10. Wood,D.D., Ihrie,E.J., Dinarello,C.A. and Cohen,P.L. (1983) *Arthritis Rheum.*, **26**, 975.
11. Nishijima,H., Weil,M.H., Shubin,H. and Cavanilles,J. (1973) *Medicine*, **52**, 287.
12. Beutler,B., Mahoney,J., Le Trang,N., Pekala,P. and Cerami,A. (1985) *J. Exp. Med.*, **161**, 984.
13. Beutler,B., Milsark,I.W. and Cerami,A. (1985) *J. Immunol.*, **135**, 3972.
14. Tracey,K.J., Beutler,B., Lowry,S.F., Merryweather,J., Wolpe,S., Milsark,I.W., Hariri,R.J., Fahey,T.J., Zentella,A., Albert,J.D., Shires,G.T. and Cerami,A. (1986) *Science*, **234**, 470.
15. Mahoney,J.R., Beutler,B.A., Le Trang,N., Vine,W., Ikeda,Y., Kawakami,M. and Cerami,A. (1985) *J. Immunol.*, **134**, 1673.
16. Oliff,A., Defeo-Jones,D., Boyer,M., Martinez,D., Kiefer,D., Vuocolo,G., Wolfe,A. and Socher,S.H. (1987) *Cell*, **50**, 155.
17. Young,D.C., Wagner,K. and Griffin,J.D. (1987) *J. Clin. Invest.*, **79**, 100.
18. Griffin,J.D., Young,D., Herrman,F., Wiper,D., Wagner,K. and Sabbath,K.D. (1986) *Blood*, **67**, 1448.
19. Ymer,S., Tucker,Q.J., Sanderson,C.J., Hapel,A.J., Campbell,H.D. and Young,J.G. (1985) *Nature*, **317**, 255.
20. Arya,S.K. and Gallo,R.C. (1987) *Lymphokines*, **13**, 35.
21. Mitelman,F. (1985) In Sanberg,A.A. (ed.), *Progress and Topics in Cytogenetics.* Alan R.Liss, New York, Vol. 5, p. 107.
22. Liao,Z., Grimshaw,R.S. and Rosenstreich,D. (1984) *J. Exp. Med.* **159**, 126.
23. Tiku,K., Tiku,M.L., Liu,S. and Skosey,J.L. (1986) *J. Immunol.*, **136**, 3686.
24. Donahue,R.E., Wang,E.A., Stone,D.K., Kaman,R., Wong,G.G., Sehgal,P.K., Nathan,D.G. and Clark,S.C. (1986) *Nature*, **321**, 872.
25. Nienhuis,A.W., Donahue,R.E., Karisson,S., Clark,S.C., Agricola,B., Antinoff,N., Pierce,J.E., Turner,P., Anderson,W.F. and Nathan,D.G. (1987) *J. Clin. Invest.*, **80**, 573.
26. Metcalf,D., Begley,C.G., Johnson,G.R., Nicola,N.A., Lopez,A.F. and Williamson, D.J. (1986) *Blood*, **68**, 46.
27. Welte,K., Bonilla,M.A., Gillio,A.P., Boone,T.C., Potter,G.K., Gabrilove,J.L., Moore,M.A.S., O'Reilly,R.J. and Souza,L.M. (1987) *J. Exp. Med.*, **165**, 941.
28. Nathan,C.F., Kaplan,G., Levis,W.R., Nusrat,A., Witmer,M.D., Sherwin,S.A., Job,C.K., Horowitz,C.R., Steinman,R.M. and Cohn,Z.A. (1986) *N. Engl. J. Med.*, **315**, 6.
29. Grimm,E.A., Mazumder,A., Zhang,H.Z. and Rosenberg,S.A. (1982) *J. Exp. Med.*, **155**, 1823.
30. Mule,J.J., Shu,S., Schwarz,S.L. and Rosenberg,S.A. (1984) *Science*, **225**, 1487.
31. Rosenberg,S.A., Lotze,M.T., Muul,L.M., Leitman,S., Chang,A.E., Ettinghausen, S.E., Matory,Y.L., Skibber,J.M., Shiloni,E., Vetto,J.T., Seipp,C.A., Simpson,C. and Reichert,C. (1985) *N. Engl. J. Med.*, **313**, 1485.
32. Nathan,C.F., Horowitz,C.R., De la Harpe,J.D., Vadhan-Raj,S., Sherwin,S.A., Oettgen,H.F. and Krown,S.E. (1985) *Proc. Natl. Acad. Sci. USA*, **82**, 8686.

6

Endpiece: The future for lymphokines

Lymphokines are polypeptides with powerful and wide ranging effects on the immune system and beyond. In 20 years they have emerged from a restricted role as mediators of cellular immunity, to hormones with influence on other physiological systems. Thus, they not only activate and amplify cellular responses to antigens but mediate many responses to stress. Co-incident with our appreciation that immune cells are not solely involved with defence against infection has come experimental evidence that lymphokines influence many cell types in many organs where they facilitate intercellular communication. As such they are not only of interest to immunologists but also cell biologists and physiologists concerned with both homeostasis and dysfunction at the cellular and whole-animal level.

The major recent advance has been the molecular biology of lymphokines. Not only are we able to say that they are 'real' and not just an artefact of culture systems, but we are gaining knowledge about their regulation at the DNA level. It is not clear how many lymphokines there are nor how they are related although structural similarities are emerging and the predominant location to date of the genes on chromosome 5 perhaps suggests a common origin and gene duplication. It is possible that in the future a 'lymphokine super-gene family' will take its place in immunological parlance.

Gene cloning has provided the material for biological studies with homogeneous material. The apparent initial euphoria has become a little tempered by the realization that, whilst the test substance (i.e. lymphokine) is well characterized, the test system is not. Bioassays are the best available method for evaluating lymphokine concentrations and action at the moment. As outlined in Chapter 1, the fact that lymphokines can initiate complicated effects within mixed cell populations may mean that bioassays are difficult to interpret. They are notoriously subject to day-to-day and between-laboratory variation and this realization has provoked growing interest in the development of properly standardized tests for lymphokines. Other problems have emerged—recombinant lymphokines may contain endotoxin and this stimulates cells to produce interleukin 1 (IL-1) which, as seen here, is not far short of 'ubiquitin'. This means that purity is still an issue, even when using recombinant proteins. All of this suggests that recombinant materials have produced a new set of challenges for the immunologist wishing to delve into the biological activity of lymphokines.

Most of what we know about the biological activity of lymphokines has come from *in vitro* experiments. It is possible that some of these activities are rarely or never seen *in vivo*. There is a pressing need to understand how lymphokines work *in vivo* and where. Immune responses occur primarily in tissues particularly lymph nodes where T and B cells are organized into discrete areas in association with antigen-presenting cells. Which cells have receptors for lymphokines in normal and diseased tissues? Which cells make lymphokines and when? Techniques such as *in situ* hybridization could provide useful information regarding the spatial and temporal production of lymphokines during an immune response and this information is badly needed. We have little idea about the lymphokine concentrations in intercellular spaces or the distance over which a lymphokine can act and these issues are of great importance for appreciating the potential a lymphokine has for exerting its effects systemically. This in turn is important for our view of regulation of haematopoiesis during stress, cachexia and so on where it is envisaged that lymphokines produced at a distant site (where?) might influence many body systems.

At the cellular level the developing area is that of receptor structure and function. At the moment the structure of lymphokine receptors and their distribution is mostly unknown. The work is hampered because many lympho-kine receptors are expressed in very low numbers, making both development and usage of monoclonal antibodies against them difficult. However, in the next few years, the structure of the receptors as well as the lymphokines will become available, and there will be investigation of the events involved in ligand receptor triggering. How does the cell transmit a message from the outside to its genes? This question is by no means unique to those interested in lymphokines and presumably there will be much to be learnt from other systems where these interactions are now being studied.

It is of interest to ask why there seem to be different gene products which interact with the same receptor (e.g. tumour necrosis factor and lymphotoxin, IL-1α and IL-1β). Does the redundancy allow a fail-safe mechanism to ensure an immune response occurs? Do these substances actually initiate the same processes? Are the lymphokines produced under totally different conditions and therefore initiate the same response but to a different stimulus? There are no answers to these questions as yet.

The potent activity of lymphokines and their availability in recombinant form has led to a burst of studies on their therapeutic usage. The potential will probably take some time to realize as studies are needed to define the optimum non-toxic levels. Since it is likely that in many situations 'cocktails' of lymphokines may be beneficial, it will take time to assemble the information on these substances separately and together. It is tempting to speculate that at the end of the day a mixture of recombinant materials which resembles the 'natural' cellular mixture may turn out to be the most useful.

In this book I have not covered all of the lymphokines but have focused on those which have been cloned. There has been no mention of migration inhibitor factor and no mention of helper and suppressor factors as well as many other lymphokines. This does not imply that they are not extremely biologically important since the complete function of any lymphokine, cloned or uncloned, has yet to be fully established. In 1988 we can say that lymphokines have come a long way but certainly have a long way to go.

Glossary

Autocrine: referring to the way in which a cell can produce a factor which acts on itself (cf. paracrine).

Bioassay: a method for measuring molecules such as lymphokines via their physiological actions on cells or living creatures.

Cachexia: weight loss and wasting often seen in individuals carrying tumours.

CD molecules: a system of nomenclature for leukocyte surface molecules, including:
CD2, present on T cells and involved in antigen non-specific cell activation;
CD3, present on T cells associated with the antigen receptor and involved in antigen-specific cell activation;
CD4, present primarily on helper T cells and involved in class-II-restricted interactions;
CD8, present primarily on cytotoxic T cells and involved in class-I-restricted interactions;
CD16, Fc IgG receptor present on natural killer cells and neutrophils;
CD23, the low affinity Fcε receptor; and
CD25, present on antigen-activated T and B cells acting as a receptor for IL-2.

cDNA library: a set of DNA fragments prepared on mRNA templates from a particular cell type. The fragments are expanded by gene cloning.

Colony stimulating factors: a group of molecules which promote division and differentiation of particular lineages of haematopoietic cells.

Delayed type hypersensitivity: a measure of helper T cell reactivity to a particular antigen, detected as a hypersensitivity reaction developing over 24 – 96 h following application of the antigen to the skin.

Domain: a globular region of folded polypeptide. Antigen receptors and MHC molecules are formed into domains.

Edman Degradation: technique for analysing the amino acid residue sequence of a peptide.

Endotoxin: a component of Gram negative bacterial cell walls (lipopolysaccharide) which can induce B cell mitogenesis, macrophage activation, complement activation and vascular shock.

Exon: gene segment encoding protein.

Glycosylation: the process of adding carbohydrate groups.

Homology: similar in structure.

Ia: generic term for MHC class II molecules.

Interferons: a group of glycoproteins which act on cells to interfere with viral replication. They also have many other physiological and immunomodulatory effects.

Interleukins: a group of molecules involved in signalling between lymphocytes, antigen-presenting cells and other cells in the body.

Introns: gene segments between exons not encoding protein.

Lymphokines: a group of molecules produced by leukocytes (primarily T cells) and acting on other cells. This description overlaps with that of some interferons and interleukins.

Open reading frame: a stretch of DNA may be read in one of three frames depending on the alignment of the codons. An open reading frame is the one which does not contain stop codons.

Paracrine: acting on other cells (cf. autocrine).

Pleiotropy: the product of a single gene being able to produce two or more effects.

Retrovirus: a group of viruses which have RNA as their genetic material, but which reverse transcribe this into DNA which may become integrated into the host cell nuclear DNA.

Signal sequence: a short stretch of amino acids at the amino-terminus of a polypeptide which directs the translation of the remainder of the polypeptide across the membrane of the endoplasmic reticulum. It is cleaved from the mature protein.

Site-directed mutagenesis: a technique for identifying the active amino acid residues in a ligand–receptor interaction, by selectively changing individual residues and looking for alterations in the function of the protein.

Stem cells: precursor cells which undergo division and give rise to the various lineages of differentiated haematopoietic cells.

Synergy: acting or working together.

Index